Entity Resolution in the Web of Data

Synthesis Lectures on the Semantic Web

Editor

Ying Ding, *Indiana University*

Paul Groth, *Elsevier Labs*

Synthesis Lectures on the Semantic Web: Theory and Application is edited by Ying Ding of Indiana University and Paul Groth of Elsevier Labs. Whether you call it the Semantic Web, Linked Data, or Web 3.0, a new generation of Web technologies is offering major advances in the evolution of the World Wide Web. As the first generation of this technology transitions out of the laboratory, new research is exploring how the growing Web of Data will change our world. While topics such as ontology-building and logics remain vital, new areas such as the use of semantics in Web search, the linking and use of open data on the Web, and future applications that will be supported by these technologies are becoming important research areas in their own right. Whether they be scientists, engineers or practitioners, Web users increasingly need to understand not just the new technologies of the Semantic Web, but to understand the principles by which those technologies work, and the best practices for assembling systems that integrate the different languages, resources, and functionalities that will be important in keeping the Web the rapidly expanding, and constantly changing, information space that has changed our lives.

Topics to be included:

- Semantic Web Principles from linked-data to ontology design

- Key Semantic Web technologies and algorithms

- Semantic Search and language technologies

- The Emerging "Web of Data" and its use in industry, government and university applications

- Trust, Social networking and collaboration technologies for the Semantic Web

- The economics of Semantic Web application adoption and use

- Publishing and Science on the Semantic Web

- Semantic Web in health care and life sciences

Entity Resolution in the Web of Data

Vassilis Christophides, Vasilis Efthymiou, and Kostas Stefanidis

ISBN: 978-3-031-79467-4 paperback
ISBN: 978-3-031-79468-1 ebook

DOI 10.1007/978-3-031-79468-1

A Publication in the Springer series
SYNTHESIS LECTURES ON THE SEMANTIC WEB

Lecture #13
Series Editors: Ying Ding, *Indiana University*
                Paul Groth, *Elsevier Labs*
Founding Editor Emeritus: James Hendler, *Rensselaer Polytechnic Institute*
Series ISSN
Print 2160-4711   Electronic 2160-472X

Entity Resolution
in the Web of Data

Vassilis Christophides
University of Crete, Greece
INRIA, France

Vasilis Efthymiou
University of Crete, Greece
ICS-FORTH, Greece

Kostas Stefanidis
ICS-FORTH, Greece

SYNTHESIS LECTURES ON THE SEMANTIC WEB #13

ABSTRACT

In recent years, several knowledge bases have been built to enable large-scale knowledge sharing, but also an entity-centric Web search, mixing both structured data and text querying. These knowledge bases offer machine-readable descriptions of real-world entities, e.g., persons, places, published on the Web as Linked Data. However, due to the different information extraction tools and curation policies employed by knowledge bases, multiple, *complementary* and sometimes *conflicting* descriptions of the same real-world entities may be provided. Entity resolution aims to identify different descriptions that refer to the same entity appearing either within or across knowledge bases.

The objective of this book is to present the *new entity resolution challenges* stemming from the *openness* of the Web of data in describing entities by an unbounded number of knowledge bases, the *semantic and structural diversity* of the descriptions provided across domains even for the same real-world entities, as well as the *autonomy* of knowledge bases in terms of adopted processes for creating and curating entity descriptions. The *scale*, *diversity*, and *graph structuring* of entity descriptions in the Web of data essentially challenge how two descriptions can be effectively compared for similarity, but also how resolution algorithms can efficiently avoid examining pairwise all descriptions.

The book covers a wide spectrum of entity resolution issues at the Web scale, including basic concepts and data structures, main resolution tasks and workflows, as well as state-of-the-art algorithmic techniques and experimental trade-offs.

KEYWORDS

entity resolution, Web of data

Contents

List of Figures

List of Tables

Preface

Over the past decade, numerous knowledge bases (KBs) have been built to power a new generation of Web applications that provide *entity-centric search* and *recommendation services*. These KBs offer comprehensive, machine-readable descriptions of a large variety of real-world entities (e.g., persons, places, products, events) published on the Web as Linked Data (LD). Even when derived from the same data source (e.g., a Wikipedia entry), KBs such as DBpedia, YAGO2, or Freebase may provide multiple, non-identical descriptions for the same real-world entities. This is due to the different information extraction tools and curation policies employed by KBs, resulting in *complementary* and sometimes *conflicting* entity descriptions. Entity resolution (ER) aims to identify different descriptions that refer to the same real-world entity, and emerges as a central data-processing task for an *entity-centric organization* of Web data. ER is needed to enrich interlinking of data elements describing entities, even by third parties, so that the Web of data can be accessed by machines as a *global data space* using standard languages, such as SPARQL. ER can also facilitate an automated KB construction by integrating entity descriptions from legacy KBs with Web content published as HTML documents.

ER has attracted significant attention from many researchers in information systems, database, and machine-learning communities. The objective of this lecture is to present the *new ER challenges* stemming from the Web *openness* in describing, by an unbounded number of KBs, a multitude of entity types across domains, as well as the *high heterogeneity* (semantic and structural) of descriptions, even for the same types of entities. The *scale*, *diversity*, and *graph structuring* of entity descriptions published according to the LD paradigm challenge the core ER tasks, namely, (i) how descriptions can be *effectively compared for similarity* and (ii) how resolution algorithms can *efficiently filter the candidate pairs* of descriptions that need to be compared.

In a multi-type and large-scale entity resolution, we need to examine whether two entity descriptions are *somehow* (or near) *similar* without resorting to domain-specific similarity functions and/or mapping rules. Furthermore, the resolution of some entity descriptions might influence the resolution of other neighborhood descriptions. This setting clearly goes beyond deduplication (or record linkage) of collections of descriptions, usually referring to a single entity type, that slightly differ only in their attribute values. It essentially requires leveraging similarity of descriptions both on their *content* and *structure*. It also forces us to revisit traditional ER workflows consisting of separate *indexing* (for pruning the number of candidate pairs) and *matching* (for resolving entity descriptions) phases.

This Synthesis lecture is intended to provide a starting point for researchers, students, and developers who are interested in a global view of the ER problem in the Web of data. Throughout the lecture, we present the basic concepts and resolution workflows, as well as state-of-the-art indexing and matching techniques. We additionally survey new ER execution strategies (such as parallel/distributed and progressive strategies) to resolve, under specific efficiency or effectiveness constraints, very large collections of entity descriptions, eventually arriving in streams. We made an effort to define in a self-contained way the similarity measures and data structures involved in various algorithms along with representative examples. We finally provide an experimental evaluation of a large part of the presented techniques and explain the involved trade-offs for real KBs in the Web of data. Much of the material presented in this lecture has been used in graduate courses taught at the University of Crete, as well as in two recent tutorials at CIKM'13 and WWW'14.

Since ER is a specialized problem of Data Integration, our Synthesis lecture provides complementary material with other books in this research area. [Doan et al., 2012] focuses on models, languages, and architectures for Data Integration systems, as well as on techniques for rewriting and processing queries on top. It also covers machine learning techniques for inferring mappings/matchings between heterogeneous relational and Web data. [Dong and Srivastava, 2015] stresses the Data Integration challenges in the Big Data era. In particular, it details how well-known ER algorithms can benefit from parallel and distributed implementations, aiming to reduce the overall execution time of the entire ER process. The book also considers schema alignment, as well as techniques for linking text snippets with embedded attributes, to structured records. Record Linkage techniques and Deduplication techniques for traditional Data Warehouse settings have been the subject of numerous surveys and books, such as [Naumann and Herschel, 2010] and [Christen, 2012]. Finally readers are referred to [Abiteboul et al., 2011] for a comprehensive overview of languages and technologies involved in Web Data Management.

Vassilis Christophides, Vasilis Efthymiou, and Kostas Stefanidis
June 2015

Acknowledgments

Several people provided valuable support during the preparation of this book, without whose help the project could not have been satisfactorily and timely completed. We warmly thank Ying Ding and Paul Groth for inviting us to write this book and Michael Morgan for managing the entire publication process. Special thanks also go to Christian Bizer for constructive feedback during the review process of the book. We would also like to acknowledge our many collaborators who have influenced our thoughts and our understanding of this research area over the years, and the following projects for their support in our research efforts: EU FP7-ICT-2011-9 DIACHRON (Managing the Evolution and Preservation of the Data Web), EU FP7-PEOPLE-2013-IRSES SemData (Semantic Data Management), EU FP7-ICT-318552 IdeaGarden (An Interactive Learning Environment Fostering Creativity), and LoDGoV (Generate, Manage, Preserve, Share, and Protect Resources in the Web of Data) of the Research Programme ARISTEIA (EXCELLENCE), GSRT, Ministry of Education, Greece, and the European Regional Development Fund. Finally, we would like to thank the ~okeanos GRNET cloud service that is used in our experimental evaluation.

Vassilis Christophides, Vasilis Efthymiou, and Kostas Stefanidis
June 2015

CHAPTER 1

Web of Data: Describing and Linking Entities

The Web bears the potential of being a universal source of knowledge used to answer questions, retrieve facts, solve problems, or create new knowledge. Many major scientific discoveries have been made possible by recognizing the connections across domains or by integrating insights from several sources [Gruber, 2008]. This process requires accessing deep insights involving the actual subject matter of scientific domains that clearly goes beyond simple associations of words supported by the existing Web infrastructure. The emerging *Web of data* aims to support a global data infrastructure, in which things of the real world (i.e., *entities*) are described on the Web by data rather than documents. This generation of the Web infrastructure is expected to transform the way structured information is exploited across domains at a large scale, and thus plays the role of a key-enabling technology for the forthcoming data-driven economy.[1] The Web of data has been proposed as a stripped-down version of the W3C specifications for the Semantic Web [Heath and Bizer, 2011] and has been boosted by the proliferation of scientific datasets made available worldwide according to the *fourth paradigm of science* [Hey et al., 2009], as well as of high-quality open encyclopedias, like Wikipedia.[2]

An increasing number of government organizations, local bodies, private companies, scientific or citizen communities are currently describing a great variety of real-world entities (e.g., persons, places, products, events) as *Linked Data* (LD). LD refers to the recent W3C efforts[3] for a unifying, machine-readable data representation infrastructure, enabling us to semantically access and interlink heterogeneous sources at the data level, no matter what the structure and the semantics of the data is, who created it, or where it comes from. The core idea is to use HTTP URIs to identify arbitrary real-world entities. Whenever a Web client resolves one of these URIs, the corresponding Web server provides the description of the identified entity under the form of a collection of RDF triples,[4] i.e., *subject-predicate-object* facts. These datasets may contain links to entities described in other Web servers. Links are also expressed in the RDF syntax, in which the subject of the triple is a URI in the namespace of one server, and the object of the triple is a URI in the namespace of the other. The predicate URI of the triple determines the type of the link. Whenever an application resolves the URI of a predicate, the corresponding server responds with

---

[1]http://ec.europa.eu/digital-agenda/en/toward-thriving-data-driven-economy
[2]https://www.wikipedia.org
[3]http://www.w3.org/DesignIssues/LinkedData.html
[4]http://www.w3.org/RDF

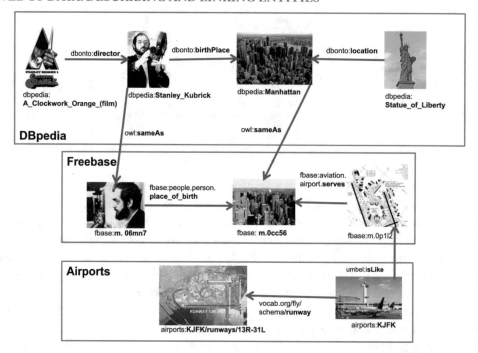

Figure 1.1: A part of the Web of data, extracted from three knowledge bases.

the definition of the link type in an RDF Schema (RDFS[5]) or Web Ontology Language (OWL[6]) syntax. Such schema descriptions can, in turn, contain links to semantic constructs (e.g., properties, classes) defined in other vocabularies [Cheng and Qu, 2013]. Exhibiting a higher degree of interoperability than documents and ease of reuse both by humans and machines, LD emerges as a prominent paradigm for publishing structured information worldwide.

Example 1.1   Consider, for example, the entities of Figure 1.1, whose descriptions are published by three knowledge bases: DBpedia,[7] Freebase[8] and Airports.[9] In this example, DBpedia describes four entities, namely the movie *A Clockwork Orange*, its director *Stanley Kubrick*, along with his birth place *Manhattan*, and the *Statue of Liberty*, which is nearby located. Freebase provides an alternative description for *Stanley Kubrick* and *Manhattan*, while it additionally describes the *John F Kennedy (JFK) International Airport* serving New York City. Another description of JFK is published by the Airports base along with one of *JFK runways*. The facts employed to describe these entities in the three knowledge bases are given as collections of RDF triples in Ta-

---

[5]http://www.w3.org/TR/rdf-schema
[6]http://www.w3.org/TR/owl2-overview
[7]http://dbpedia.org
[8]https://www.freebase.com
[9]The "airports" dataset on https://archive.org/details/kasabi

ble 1.1: each block of triples with a common subject describes an entity. You can easily observe the graph-based structure of entity descriptions expressed in RDF. Nodes of this graph represent the URLs of entities (e.g., *dbpedia:A_Clockwork_Orange_(film)*), while edges represent the different types of relationships (i.e., links) that stand among the described entities. For instance, properties like *dbonto:director*, *dbonto:birthPlace* or *dbonto:location* (having as subject and object URLs from the same namespace) relate the descriptions of different types of entities within DBpedia, while *owl:sameAs* (having as subject and object URLs from different namespaces) indicate somehow similar descriptions of the same real-world entity across different knowledge bases.

Table 1.1: The RDF triples of the entities appearing on Figure 1.1

KB	subject	predicate	object
DBpedia	dbpedia:A_Clockwork_Orange_(film)	dbonto:director	dbpedia:Stanley_Kubrick
	dbpedia:A_Clockwork_Orange_(film)	dbonto:Work/runtime	"136"
	dbpedia:A_Clockwork_Orange_(film)	rdfs:label	"A Clockwork Orange (film)"
	dbpedia:A_Clockwork_Orange_(film)	foaf:name	"A Clockwork Orange"
	dbpedia:Stanley_Kubrick	dbonto:birthPlace	dbpedia:Manhattan
	dbpedia:Stanley_Kubrick	owl:sameAs	fbase:m.06mn7
	dbpedia:Stanley_Kubrick	rdf:type	foaf:Person
	dbpedia:Stanley_Kubrick	rdf:type	yago:AmericanFilmDirectors
	dbpedia:Stanley_Kubrick	rdf:type	yago:AmateurChessPlayers
	dbpedia:Manhattan	owl:sameAs	fbase:m.0cc56
	dbpedia:Manhattan	rdf:type	yago:IslandsOfTheHudsonRiver
	dbpedia:Statue_of_Liberty	dbonto:location	dbpedia:Manhattan
Freebase	fbase:m.06mn7	fbase:type.object.name	"Stanley Kubrick"
	fbase:m.06mn7	fbase:people.person.place_of_birth	fbase:m.0cc56
	fbase:m.06mn7	fbase:people.person.parents	fbase:m.02g68r_
	fbase:m.06mn7	fbase:people.person.parents	fbase:m.02g656g
	fbase:m.0p1l2	fbase:aviation.airport.serves	fbase:m.0cc56
Airports	airports:KFJK	foaf:name	"John F Kennedy International Airport"
	airports:KFJK	foaf:homepage	http://www.panynj.gov/airports/jfk.html
	airports:KFJK	umbel:isLike	fbase:m.0p1l2
	airports:KFJK	http://vocab.org/fly/schema/runway&airports:KJFK/runways/13R-31L	

Web Knowledge Bases

Comprehensive, machine-readable descriptions of real-world entities are hosted in *knowledge bases* (KB). Entities in a KB are described as instances using concepts and relationships among them. These semantic modeling constructs are flexibly represented nowadays in the RDF syntax (see Table 1.1) overcoming the schema rigidity of traditional databases [Christophides, 2009].

Over the past decade, numerous KBs have been built to power a new generation of Web applications, such as *entity-centric search* [Balog et al., 2010a,b, Blanco et al., 2011, Haas et al., 2011, Kitsos et al., 2014, Lin et al., 2012] and *recommendations* [Blanco et al., 2013, Miliaraki et al., 2015, Yu et al., 2014]. Such KBs can be *domain-specific*, describing specific types of entities from a well-defined domain of interest, as for example, movie and music production, scientific publications, social and economic statistics and life sciences, or be *cross-domain*, containing encyclopedic knowledge for a variety of entity types, as for instance, DBpedia [Auer et al., 2007], YAGO2 [Hoffart et al., 2013], and Freebase [Bollacker et al., 2008]. Traditionally, KBs are manually crafted by a dedicated team of knowledge engineers, such as the pioneering projects Wordnet.[10] and Cyc.[11] With the explosion of the Web, more and more KBs are built from existing Web

[10]http://wordnet.princeton.edu
[11]http://www.cyc.com

content using information extraction tools [Doan et al., 2009]. Such an automated approach offers an unprecedented opportunity to scale-up KBs construction and leverage existing knowledge published in HTML documents [Hovy et al., 2013]. More precisely, we can further distinguish between [Dong et al., 2014a]:

- KBs derived from well-structured textual entries of high-quality sources, like Wikipedia infoboxes:[12] YAGO2, DBpedia, and Freebase,

- KBs derived from schemaless information extraction techniques applied over the entire Web: Reverb [Fader et al., 2011], OLLIE [Mausam et al., 2012], Microsoft Satori, Google Knowledge Graph,[13] and Knowledge Vault [Dong et al., 2014a],

- KBs derived from domain-specific Web pages using a specific ontology: NELL [Carlson et al., 2010], PROSPERA [Nakashole et al., 2011], DeepDive [Niu et al., 2012], and Facebook Entity Graph.[14]

Several questions naturally arise regarding the characteristics of the entity descriptions published by a KB on the Web. For example,

- *how many entities* are described and by *how many facts* (aka triples)?

- *what entity types* (aka classes) are covered, and *what property vocabularies* are employed to describe them?

- *which semantic relationships* (e.g., equivalence, relatedness) stand between the entities described within or across KBs?

These questions are essential in order to assess various aspects of KB data quality [Zaveri et al., 2014] and be able to choose the most suitable KBs for the needs of a specific application. In particular, they allow us to understand extrinsic characteristics of entity descriptions, such as *coverage, completeness,* and *relevance,* but also intrinsic, such as *accuracy, consistency,* and *conciseness* [Kontokostas et al., 2014]. Compared to data warehouses, the new data quality challenges stem from the Web *openness* in describing by an unbounded number of KBs a multitude of entity types across domains, as well as the *high heterogeneity* (semantic and structural) of descriptions even for the same types of entities. As a matter of fact, the *autonomy* of KBs in terms of adopted processes for creating and curating entity descriptions [Deshpande et al., 2013] results to complementary (and sometimes conflicting) descriptions even for the same real-world entities (when they evolve). The ability to reconcile multiple descriptions, within or across KBs, which refer to the same real-world entity, is crucial in order to foster an *entity-centric* organization of Web data [Bouquet et al., 2007, Dalvi et al., 2009, 2012, Miklós et al., 2010]. In particular, it will enable

---

[12]http://en.wikipedia.org/wiki/Template:Infobox
[13]http://www.google.com/insidesearch/features/search/knowledge.html
[14]https://www.facebook.com/notes/facebook-engineering/under-the-hood-the-entities-graph/
   10151490531588920

Table 1.2: Comparison of knowledge bases

Knowledge base	# Entities	# Classes	# RDF triples	# Properties
YAGO2	10M	350K	120M	100
DBpedia (en)	4.58M	685	583M	2,795
Freebase	46.3M	1.5K	2.67B	4K
Knowledge Graph [Dong et al., 2014a]	600M	15K	20B	35K
Knowledge Vault [Dong et al., 2014a]	45M	1.1K	1.6B	4.5K

interlinking of entity descriptions even by third-parties.[15] This clearly requires an understanding of the relationships among entity descriptions that goes beyond *high similarity* studied in traditional deduplication and cleaning problems [Christen, 2012, Naumann and Herschel, 2010]. We essentially need to explore entity descriptions that are *somehow* (or near) *similar* [Broder, 2000, He, 2014, Papadias, 2009] without always being able to merge related entity descriptions in a KB and thus improve its quality.[16]

In the sequel, we will take a closer look at KBs that are derived from Wikipedia, using information extraction tools and occasionally involving human volunteers. Table 1.2 summarizes their main characteristics. Only the English version of DBpedia,[17] in 2014, describes 4.58M entities using 583M triples, including 1,445,000 persons, 735,000 places, 411,000 creative works, 241,000 organizations, 251,000 species, and 6,000 diseases. All versions together (125 languages) describe 38.3M entities, out of which 23.8M are localized descriptions of entities that also exist in the English version. The full version consists of 3B triples out of which 583M were extracted from the English edition of Wikipedia and 2.46B were extracted from other language editions. It features 25.2M links to images and 29.8M links to external Web pages. Moreover, it is connected to other KBs by around 50M links. Apart from its own 685 classes that DBpedia defines to assign semantic types to its entities (e.g., Person, Place), it also contains 80.9M links to Wikipedia categories and 41.2M links to more fine-grained YAGO classes (e.g., AmateurChess-Players, IslandsOfTheHudsonRiver). Besides Wikipedia, YAGO2 also exploits knowledge from WordNet[18] and GeoNames.[19] It currently describes[20] more than 10M entities using 120M RDF triples employing 100 properties from 100 languages. It combines the clean taxonomy of Word-Net with the richness of the Wikipedia category system, assigning the entities to more than 350K classes. Moreover, it is anchored in time and space, since many of its entities and triples are associated with a temporal and spatial dimension. The accuracy of YAGO2 has been manually assessed

---

[15]For instance, the sameas.org service provides co-references of the same entities between different KBs that have been manually collected.

[16]Entity interlinking shares similar motivations with the approach of consistent query answering over inconsistent databases [Bertossi, 2011].

[17]http://wiki.dbpedia.org/Datasets

[18]http://wordnet.princeton.edu

[19]http://www.geonames.org

[20]http://www.mpi-inf.mpg.de/departments/databases-and-information-systems/research/yago-naga/yago/

to reach 95%. To enhance data completeness, Freebase allows structured descriptions of entities to be contributed by volunteers (after approval) in the spirit of the original Wikipedia. It currently describes[21] 46.3M entities in 1,500 classes, and contains 2.67B triples for 4K properties, where only 637M triples are non-redundant. Still, 75% of people descriptions in Freebase have no known nationality, 91% have no known education, 68% have no known profession, and 71% have no known place of birth [Bordes and Gabrilovich, 2014]. A recent Google endeavor, the Knowledge Graph KB fuses entity descriptions from several KBs, including Freebase and the CIA World Factbook.[22] It currently contains about 600M entities in 15K classes and more than 20B triples for 35K properties.

Finally, Google's Knowledge Vault pushes forward the automated KB construction process, by combining descriptions of entities published in existing KBs, like Freebase, with facts extracted from free text, DOM trees, Web tables,[23] or human annotations of Web pages. To enhance trustworthiness, each triple is associated with a confidence score, representing the probability that the Knowledge Vault "believes" the triple is correct. Currently, it contains 1.6B triples, out of which 324M have a confidence of 0.7 or higher, and 271M have a confidence of 0.9 or higher. Moreover, it describes 45M entities in 1,100 classes.

Semantic Annotations of Web pages

Besides KBs, machine-readable descriptions of real-world entities are also published in the Web of data as semantic annotations of HTML pages under a variety of markup formats (Microdata,[24] RDFa,[25] Microformats[26]). The embedded semi-structured data is crawled together with the unstructured HTML pages by search engines (e.g., Google,[27] Yahoo!,[28] Yandex,[29] and Bing[30]) in order to support entity-centric Web applications involving search (e.g., e-shopping, news stories) or not (e.g., Google Now, Pinterest) [Balog et al., 2010a,b, Blanco et al., 2011, Haas et al., 2011, Lin et al., 2012]. An entity describes a "thing" (i.e., a building, person, or event) that is recognizable as such and will be described using certain properties.

Until recently, only the big search engine companies had access to large quantities of such entity descriptions as they were the only ones possessing large Web crawls. With the advent of the Common Crawl[31] non-profit foundation for Web crawling, a subset of the pages from each website, along with a subset of their Microdata, RDFa, Microformats annotations, are regularly made publicly available. Moreover, the Web Data Commons[32] project regularly extracts from the Com-

---

[21]https://developers.google.com/freebase/data
[22]https://www.cia.gov/library/publications/the-world-factbook
[23]More on exploiting valuable information from Web tables can be found in [Bizer, 2014, Cafarella et al., 2008].
[24]http://www.w3.org/TR/microdata
[25]http://www.w3.org/standards/techs/rdfa
[26]http://microformats.org
[27]https://developers.google.com/structured-data
[28]https://developer.yahoo.com/searchmonkey/siteowner.html
[29]https://webmaster.yandex.com/microtest.xml
[30]http://www.bing.com/webmaster/help/marking-up-your-site-with-structured-data-3a93e731
[31]http://commoncrawl.org
[32]http://www.webdatacommons.org

mon Crawl structured data (from semantic annotations, as well as Web Tables[33]), and publish them for research purposes. According to the latest extraction campaign,[34] semi-structured data were found within 620 million HTML pages out of the 2.01 billion pages contained in the crawl (30%). Altogether, the extracted datasets consist of 20.48 billion RDF quads (the fourth element represents the provenance of each triple). For example, over 90,000 e-commerce websites annotate product data, while 60,000 websites provide data about local businesses. These websites include the major players in many domains so that it can be assumed that nearly all real-world entities that exist (such as products or hotels) are covered by the extracted semi-structured data [Meusel et al., 2014]. Besides Web-scale sources of entity descriptions extracted by semantic annotations and Web tables, a number of commercial sites dispose descriptions of specialized entities, such as products (e.g., UPC,[35] GTIN13.com,[36] Smoopa,[37]) points of interest (e.g., restaurants,[38] hotels[39]) or even people (e.g., yasni[40]). These sites essentially harvest descriptions of unique Web entities, but do not make them available for public use.

It is worth mentioning that to avoid semantic discrepancies of semi-structured data embedded in Web pages, the above search engine companies have agreed on a common vocabulary[41] for describing entities across their applications [Guha, 2013]. For instance, thanks to Schema.org integrations, Google can instantly understand whether the content of a website concerns a movie, a person, a TV series, etc. Clearly, there is a continuous increase in the number of pay-level domains (PLD) adopting the Schema.org classes. Currently, 2.72 million PLDs out of the 15.68 million PLDs covered by the WebDataCommons crawl[42] provide such annotations (17%), while the adoption rates on websites that are in the Alexa top lists for e-commerce, tourism and job websites are above 60% in all three domains.[43] Given that the issue of "entities" (rather than "keywords") appeared on the radar of the search engine industry only recently, we expect an increase in this ratio in the coming years.

The quality of entity descriptions published in this fragment of the Web of data varies a lot [Meusel and Paulheim, 2015]. Most entities are only marked up with a relatively small number of general-purpose properties (e.g., name or description), leading to rather flat and abstract descriptions. As the Web is an open and unrestricted information environment, entity descriptions might be outdated or simply wrong [Meusel et al., 2014] due to mistakes in the underlying

---

[33]HTML tables on the Web are used not only for layout purposes, but also contain high-quality entity descriptions whose types and properties are understood by analyzing tables' content and surrounding textual information [Balakrishnan et al., 2015]. Readers are refered to [Dalvi et al., 2012, Doan et al., 2009, Grishman, 2012] for recent surveys on the information extraction techniques used in this context.

[34]http://webdatacommons.org/structureddata/2014-12/stats/stats.html

[35]http://www.upcdatabase.com

[36]http://gtin13.com

[37]http://www.smoopa.com

[38]http://yelp.comandhttp://urbanspoon.com

[39]http://www.booking.comandhttp://www.tripadvisor.com

[40]http://www.yasni.com

[41]http://schema.org

[42]http://www.webdatacommons.org/structureddata/2014-12/stats/stats.html

[43]http://www.slideshare.net/bizer/bizer-bis2015web-asglobaldataspace(Slide24ff)

database or to noise in the extraction process. In order to improve various aspects of data quality [Kontokostas et al., 2014, Zaveri et al., 2014], descriptions from several sources need to be resolved in order to identify and fuse those that refer to the same real-world entities [Dong and Naumann, 2009, Dong et al., 2014b]. We believe that entity resolution in this context can benefit a lot from deduplication techniques proposed in a data warehouse setting. On the contrary, discovering and selecting the most appropriate number of Web sources to support a given coverage threshold is a challenging task. As reported in the literature, even for domains with strong head aggregators, we need to go to the long tail of Web sources to build a complete database of entities [Dalvi et al., 2012].

Entity Interlinking

Following the Linked Open Data (LOD) guidelines, data publishers are encouraged to describe and interlink real-world entities using the RDF data model. Entities are named using HTTP URIs, so that people can access their descriptions on the Web, while more entities can be discovered by following different types of semantic relationships, encoded as links between entities. Interlinking is conducted by a variety of participants including KB providers, KB consumers, and third parties, and enables one to move through a potentially endless Web of related entity descriptions. The so-constructed Web of data is usually represented by the *LOD cloud*,[44] in which nodes are KBs (aka RDF datasets) and edges are links crossing KBs.[45] The 2014 version of the LOD cloud, shown in Figure 1.2, consists of 1014 KBs from eight broad domains (e.g., cross-domain, media, geographic, publication, social networking, life sciences) with more than 60B triples, whose predicates are defined in 649 distinct vocabularies.[46]

In general, although very large KBs are being added to the LOD cloud on a regular basis (e.g., Linked TCGA [Saleem et al., 2013]), they are only sparsely interlinked. Recent studies show that 44% of the LOD KBs are not at all linked [Schmachtenberg et al., 2014]. If we take a closer look at the latest LOD version, shown in Figure 1.2, only 56.11% of the KBs link to at least another KB, while 17.36% of them link to only one other KB. As a matter of fact, the distribution of links across KBs is heavily skewed in the LOD cloud.[47] Figure 1.3 shows the distribution of the in- and out-degrees of the KBs belonging to one of the eight domains. Sparsely interlinked KBs appear in the periphery of the LOD cloud (e.g., Open Food Facts, Bio2RDF), while heavily interlinked ones lie at the center (e.g., DBpedia, GeoNames, FOAF). Unsurprisingly, encyclopedic KBs, such as DBpedia, or widely used georeferencing KBs, such as GeoNames, are interlinked with the largest number of KBs both from the LOD center and the periphery.

Table 1.3 summarizes the top three properties that are used by RDF links across KBs in each of eight domains. In their majority, the links capture equivalence relationships (e.g., *owl:sameAs*) between partial, overlapping descriptions of the same real-world entities in different KBs. General

---

[44]http://lod-cloud.net
[45]Actually, edges indicate the existence of at least 50 links between two KBs.
[46]http://linkeddatacatalog.dws.informatik.uni-mannheim.de/state
[47]http://linkeddata.few.vu.nl/wod_analysis

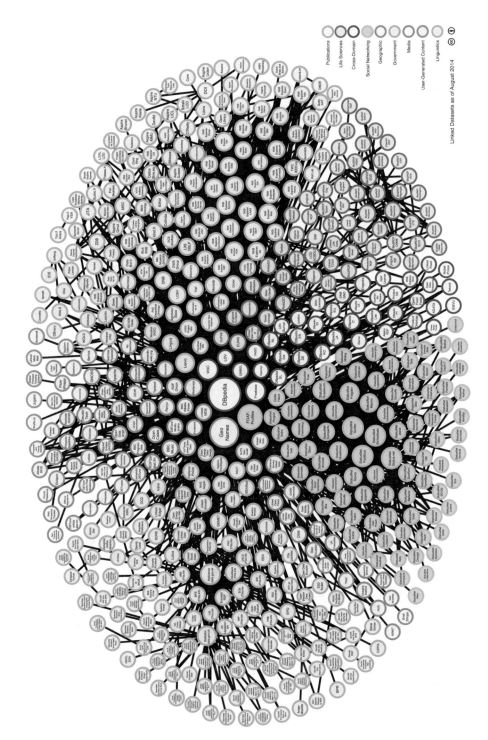

Figure 1.2: The Linked Open Data Cloud of Web KBs.[a]

[a]Linking Open Data cloud diagram 2014, by Max Schmachtenberg, Christian Bizer, Anja Jentzsch and Richard Cyganiak. `http://lod-cloud.net/`.

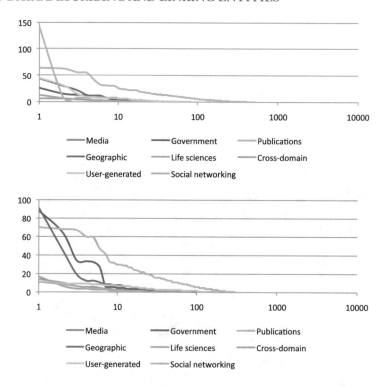

Figure 1.3: In- (top) and out-degree (bottom) distributions of different categories of datasets.

relatedness (e.g., *rdfs:seeAlso*) is the second most popular type of links, while in the social network-ing domain, social relations among persons (e.g., *foaf:knows*) is the most widely used linking type. Taking a closer look at the *owl:sameAs* property, Table 1.4 shows the top 10 KBs, based on the number of incoming *owl:sameAs* links.[48]

Due to the different participants and tools [Ferrara et al., 2013], the quality of links dis-covered across KBs may also vary [Guéret et al., 2011, Papaleo et al., 2014]. In this task, ei-ther general-purpose (e.g., Silk [Volz et al., 2009], ODD Linker [Hassanzadeh et al., 2009], LIMES [Ngomo and Auer, 2011]), or dataset-specific tools (e.g., for LinkedGeoData [Auer et al., 2009], LinkedMDB [Hassanzadeh and Consens, 2008], Music [Raimond et al., 2008]) can be used. The techniques underlying link discovery tools can be further refined as *rule-based* or *learning-based*. Rule-based methods (e.g., [Volz et al., 2009]) employ hand-crafted correspon-dence rules among entity descriptions. The creation of such rules is labor-intensive and difficult to generalize across domains. Learning-based methods (e.g., [Isele and Bizer, 2012, 2013]) try to learn such complex correspondence rules, based on a given training set of labeled examples,

[48]http://planet-data.eu/sites/default/files/D4.5.pdf

Table 1.3: Top-3 properties used by RDF links within each topical domain in the 2014 LOD cloud

Category	Property	Usage	Category	Property	Usage
social networking	foaf:knows	60.27%	user-generated content	owl:sameAs	53.13%
	foaf:based-near	35.69%		rdfs:seeAlso	21.88%
	sioc:follows	34.34%		dct:source	18.75%
life sciences	owl:sameAs	52.17%	geographic	owl:sameAs	64.29%
	rdfs:seeAlso	43.48%		skos:exactMatch	21.43%
	dct:creator	21.74%		skos:closeMatch	21.43%
publications	owl:sameAs	32.20%	media	owl:sameAs	81.25%
	dct:language	25.42%		rdfs:seeAlso	18.75%
	rdfs:seeAlso	23.73%		foaf:based-near	18.75%
government	dct:publisher	47.57%	cross-domain	owl:sameAs	80.00%
	dct:spatial	30.10%		rdfs:seeAlso	52.00%
	owl:sameAs	24.27%		dct:creator	20.00%

Table 1.4: Top 10 KBs based on their number of incoming *owl:sameAs* links

KB	# KBs linked with an *owl:sameAs* link
dbpedia.org	75
freebase.com	31
semanticweb.org	24
l3s.de	24
geonames.org	21
purl.org	16
fu-berlin.de	15
identi.ca	14
dbtune.org	13
w3.org	12

or on employing a manual verification phase. However, obtaining this set of examples is often hard when the number of KBs becomes large. In a nutshell, existing link discovery tools (for an exhaustive review readers are referred to [Nentwig et al., 2015]) are *source-centric*, i.e., they can be effectively deployed and tuned only for specific datasets. However, the range of KBs which are today published in the LOD cloud (see Figure 1.2), as well as the emergence of domain-independent techniques that holistically extract information from the entire Web [Dalvi et al., 2012] require novel entity resolution techniques at the Web-scale involving multiple, heterogeneous, and sometimes structured-related entity types.

Entity Resolution Challenges in the Web of Data

Entity descriptions in the Web of data are essentially characterised by:

- **Large Scale.** A *massive volume* of descriptions (in the order of millions) related to a *wide range of entity types* (in the order of thousands) is not ceasing to be published by an increasing number of KBs (in the order of hundreds). For example, in 2014, without counting the 20.48 billion triples extracted from the semantic annotations of Web pages [Meusel et al., 2014], more than 60B triples of entity descriptions were published in the Web of data by 1,014 open KBs (see Figure 1.2), while corporate KBs alone already qualify as big data, such as the Knowledge Graph that describes 600M entities using 20B triples.

- **High heterogeneity.** The descriptions contained in these KBs present a high degree of *semantic* and *structural* diversity even when they concern the same entity types.

  - **Semantic discrepancies.** Although the LOD publication guidelines[49] strongly suggest the reuse of entity names rather than the invention of new ones, in practice, this is not the case [Hogan et al., 2012]. Moreover, rather than using a unique classification schema, entity descriptions are annotated simultaneously with *several semantic types*, not necessarily from the same vocabulary [Tonon et al., 2013], indicating various facets for those entities.

    Example 1.2   In the example of Figure 1.1, DBpedia and Freebase employ *different names* for similar entities (e.g., the URI of Manhattan is *dbpedia:Manhattan* vs. *fbase:m.0cc56*)), but also for similar properties (e.g., the URI of a birth place is *dbonto:birthPlace* vs. *fbase:people.person.place_of_birth*). Moreover, DBpedia provides the following semantic types for Kubrick, which are not related via subsumption relationships:
    <dbpedia:Stanley_Kubrick, rdf:type, foaf:Person>,
    <dbpedia:Stanley_Kubrick, rdf:type, yago:AmericanFilmDirectors> and
    <dbpedia:Stanley_Kubrick, rdf:type, yago:AmateurChessPlayers>

  - **Structural discrepancies.** Even for the same semantic types of entities, quite different sets of properties can actually be used by different descriptions both in terms of types and number of occurrences [Duan et al., 2011].

    Example 1.3   To define the position of the Eiffel Tower, DBpedia uses the properties *rdf:lat* and *rdf:long*, as in the triples: <dbpedia:Eiffel_Tower, rdf:lat, 48.858223> and <dbpedia:Eiffel_Tower, rdf:long, 2.294500>. In the same KB, *rdf:lat* and *rdf:long* are not present in the description of the Statue of Liberty; a different set of properties is used instead:
    <dbpedia:Statue_of_Liberty, dbpprop:latDegrees, 40>,
    <dbpedia:Statue_of_Liberty, dbpprop:latMinutes, 41>,
    <dbpedia:Statue_of_Liberty, dbpprop:latSeconds, 21>,
    <dbpedia:Statue_of_Liberty, dbpprop:latDirection, N>,
    <dbpedia:Statue_of_Liberty, dbpprop:longDegrees, 74>,
    <dbpedia:Statue_of_Liberty, dbpprop:longMinutes, 2>,

---

[49]http://www.w3.org/DesignIssues/LinkedData.html

<dbpedia:Statue_of_Liberty, dbpprop:longSeconds, 40> and
<dbpedia:Statue_of_Liberty, dbpprop:longDirection, W>.

- **Various forms of overlaps.** Descriptions of the same real-world entity could be provided not only among different KBs, but even within the same KB. This is due to several reasons. Even if entity descriptions are derived from the same Wikipedia entry, KBs rely on different information extraction tools and curation policies leading to somehow similar descriptions of the same entity. This way, usually, descriptions provide complementary and sometimes conflicting information regarding evolving real-world entities. [Dalvi et al., 2012] investigates the redundancy of information that can be found on structured data on the Web. To do this, it uses the notion of k-coverage,[50] and shows, for instance, that one needs 5,000 sources to get 5-coverage of 90% of the restaurant phone numbers, while 10 sources is sufficient to get 1-coverage of 93% of these phone numbers. A systematic computation of the k-coverage among the KBs of the LOD cloud is more challenging, since it involves multiple entity types. A rough approximation of the overlap between entity descriptions in the Web of data is given by the number of *owl:sameAs* and more generally relatedness links depicted in Table 1.3. In general, a KB containing multiple descriptions for the same entity is called *dirty*, as opposed to *clean* KBs, which do not feature such internal redundancy.

Example 1.4 In our motivating example, DBpedia and Freebase describe separately both Stanley Kubrick and Manhattan. Freebase contains the information that Stanley Kubrick's parents are Gertrund Kubrick and Jacques Leonard Kubrick, whereas this information is not covered in DBpedia. On the other hand, at the time of writing, Freebase holds the information that Alexis Tsipras is the prime minister of Greece, while DBpedia is out-of-date, suggesting that this position is still held by Antonis Samaras. As a traditional form of duplicates, consider that both *dbpedia:Robert_Soloway* and *dbpedia:Spam_King* refer to the same individual, Robert Soloway, who was nicknamed the "Spam King," while both *dbpedia:Dichopogon_strictus* and *dbpedia:Chocolate_lily* refer to the same flower.

The *scale*, *diversity*, and *graph structuring* of entity descriptions published according to the Linked Data paradigm challenge the way descriptions can be effectively and efficiently compared to decide whether they are referring to the same real-world entity. Entity resolution at the Web scale requires novel algorithms and similarity functions that go beyond deduplication algorithms in data warehouses [Christen, 2012, Naumann and Herschel, 2010]. It is also worth noticing that ontology and instance matching algorithms [Otero-Cerdeira et al., 2015, Shvaiko and Euzenat, 2013] cannot be used in this respect. In the Web of data, the same description may instantiate several semantic types from different vocabularies, and thus seeking for correspondences between the different ontology classes and properties, to match the instances using them, is not feasible at

---

[50]k-coverage is defined as the fraction of entities in a database that are present in at least k different sources.

large scales. We believe that such multiple classification of instances amplifies structural heterogeneity of entity descriptions and really challenges even the most sophisticated algorithms, like Paris [Suchanek et al., 2011], aiming at a fruitful interplay between vocabulary (i.e., T-Box) and instance (i.e., A-Box) matching.

The Value of Entity Resolution

There is an undergoing paradigm shift in the Web from a document-centric infrastructure, where unstructured documents are interlinked with untyped hyperlinks, to an *entity-centric organization* of data [Bouquet et al., 2007, Dalvi et al., 2009, 2012, Miklós et al., 2010], in which real-word entities are described with semi-structured data in several KBs, or even as semantic annotations of HTML pages, and can be fused, or interlinked using various types of relationships. A core component of this transformation is our improved understanding of the contents of documents, replacing the shallow summarization of documents sketched out by keywords with a deeper association between documents and the entities mentioned in them [Jin et al., 2014].

In general, descriptions from several sources need to be resolved in order to identify and possibly fuse those that refer to the same real-world entities. For example, consider that a number of commercial sites dispose descriptions of specialized entities, such as products, restaurants, hotels, or people. When such descriptions are extracted by semantic annotations and Web tables, typically, they are marked up with a small number of general purpose properties. So, descriptions representing the same entity can be fused to improve various aspects of data quality [Dong and Srivastava, 2015]. In more complex cases, entity descriptions can be annotated simultaneously with different semantic types, indicating various facets of those entities. Interlinking such descriptions, by exploiting the several types of available links, helps toward identifying complementary descriptions for the same entities.

Together with other technologies to allow better understanding of searchers' intents, an entity-centric Web infrastructure enables powerful new user experiences, from search results that directly show key facts about people, places, and things, to improved refinement interfaces that allow searchers to quickly locate Web documents that mention only the specific people, places, or other things they are looking for [Jansen and Spink, 2006]. We are witnessing a new generation of Web applications that rely on entity descriptions to better serve navigational or information seeking needs of users, namely, *entity-centric search* [Balog et al., 2010a,b, Blanco et al., 2011, Lin et al., 2012] and *recommendations* [Blanco et al., 2013, Miliaraki et al., 2015, Yu et al., 2014]. The former semantically enrich the answers of keyword queries with references to entities that are mentioned in the queries,[51] while the latter also provides recommendations of related entities based on relationships explicitly encoded in a KB [Fang et al., 2011].

Example 1.5   Consider that a user knows exactly what he is looking for and would like immediate and precise answers, for instance, about *Stanley Kubrick*. To serve this navigational request

---

[51]A process known as named-entity extraction [Bizer et al., 2009, Etzioni et al., 2005, Hoffart et al., 2013] and disambiguation [Jin et al., 2014].

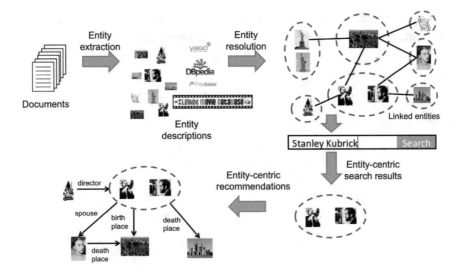

Figure 1.4: Searching and recommending entities related to "Stanley Kubrick."

according to the entity-centric paradigm, the following process would be engaged (Figure 1.4). Initially, a number of entity descriptions related to the entertainment industry (e.g., film makers) have been extracted from semantic annotations of Web pages and/or from domain specific KBs (e.g., LinkedMDB[52]) and cross-domain KBs (e.g., DBpedia, YAGO, Freebase). Such descriptions can be possibly fused or interlinked to each other. Then, the mentions of various entities in the user queries are recognized and matched to the extracted entity descriptions. For example, besides Web documents related to "Stanley Kubrick," an entity search system would enrich the answer with the descriptions of Stanley Kubrick in DBpedia and/or Freebase. To serve users willing to extend their knowledge or simply satisfy their curiosity, an entity recommender system could provide additional entities describing information of potential interest for the user. For example, consider the information that Kubrick was born in *Manhattan*, extracted from DBpedia, that he was married to Ruth Sobotka, extracted from YAGO, that he was the director of the movie *A Clockwork Orange*, extracted from LinkedMDB, and so forth.

To foster an *entity-centric* organization of Web data, it is crucial to reconcile different descriptions that refer to the same real-world entity. Entity resolution is expected to play a catalytic role for *exploratory search and discovery* [Marie and Gandon, 2014] in the Web of data. First, because it enhances fusion and interlinking of data elements describing entities, so that the Web of data can be accessed by machines as a *global data space* using standard languages, such as SPARQL. Second, because it speeds up KB construction by integrating entity descriptions from legacy KBs with semantic annotations published along with HTML pages or even HTML tables themselves.

---

[52]www.linkedmdb.org

The faster, more complete and accurate knowledge integration on the Web is, the better quality of service is offered by entity-centric search and recommendations.

CHAPTER 2

Matching and Resolving Entities

As we have seen in Chapter 1, an increasing number of real-world entities are described by a multitude of Knowledge Bases (KBs) published in the Web of data. These descriptions may provide *partial*, *overlapping*, and sometimes *contradicting* information for the same entities. Understanding how two descriptions are related is an essential task to a number of *entity-centric applications* in the Web of data (e.g., searching, browsing). Due to the open number of KBs and their autonomy in publication and curation policies, two descriptions of the same entity may significantly differ in terms of employed vocabularies and data structuring. In this chapter, we outline the main processing steps for resolving entities at the Web scale (Section 2.1) and explain how the similarity of highly heterogeneous descriptions can be defined (Section 2.2).

2.1 THE PROBLEM OF ENTITY RESOLUTION

In the Web of data, the same real-world entity may be described multiple times in various KBs (and more rarely within the same KB). These descriptions are characterised as *highly similar* or *somehow similar*. The former case corresponds to duplicate descriptions originating from the same knowledge source (e.g., Wikipedia) and copied by different KBs. The latter case corresponds to alternative descriptions of the same entity eventually covering complementary aspects (e.g., information on the work or the origins of an artist). Highly similar descriptions usually feature many common tokens in the values of common attributes, while somehow similar descriptions have significantly fewer common tokens in attributes that are not necessarily semantically related (via synonymy or subsumption). Hence, two highly similar descriptions can be compared using only their content (i.e., attribute values), while to decide whether two somehow similar descriptions refer to the same real-world entity more contextual information is needed, as for example, examining the similarity of neighborhood descriptions (i.e., linked with different types of relationships).

Example 2.1   As a toy example, consider the entity descriptions presented in Figure 2.1. DBpedia describes two movies, *Eyes Wide Shut* and *A Clockwork Orange*, their director *Stanley Kubrick* and his place of birth *Manhattan*, while Freebase provides alternative descriptions for the same four entities. When considering the tokens of the descriptions referring to *Eyes Wide Shut*, we can say that these descriptions are highly similar (i.e., they share 6 tokens), and deduce that they

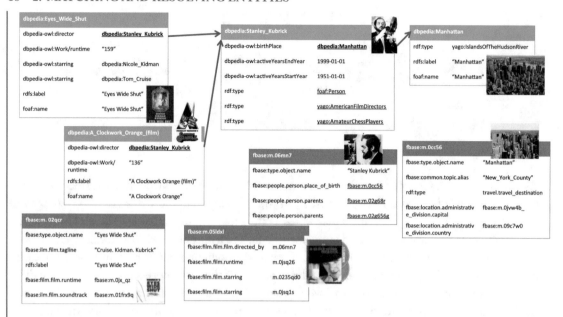

Figure 2.1: Multiple entity descriptions.

possibly stand for the same entity. However, this is not the case when comparing the somehow similar descriptions of *Stanley Kubrick*; these descriptions are also quite different with respect to the attributes used. To decide whether the descriptions of *Stanley Kubrick* in DBpedia and Freebase match, we need to consider the similarity of descriptions related to them. For instance, knowing that in DBpedia and Freebase the descriptions for *Eyes Wide Shut* stand for the same entity may help us to decide whether the related descriptions for *Stanley Kubrick* (via *movie director* relationship) in both KBs also refer to the same entity.

Finding somehow similar information has been studied in the context of near-duplicate documents detection [Broder, 2000] and nearest neighbors search [Papadias, 2009]. In the context of the Web of data, in order to compare a pair of entity descriptions we need to consider similarity of both their textual content (as in unstructured documents) and their graph-structure (as in spatio-temporal databases). Clearly, more accurate comparisons can be performed if we know the semantic relationships of attributes employed in descriptions (e.g., synonymy or subsumption), but such ontological alignment [Shvaiko and Euzenat, 2013] is not yet systematically provided at the scale of the Web of data. It is worth noticing that even for entity descriptions of the same type originating from the same KB (see Chapter 1), various sets of attributes can be employed [Duan et al., 2011], while many combinations of attributes appear only for a single description [Neumann and Moerkotte, 2011]. In general, the typically high degree of semantic heterogeneity reflected in different schemas makes ontology and schema matching in the Web of data an inherently complex task [Bellahsene et al., 2011].

Based on the locality of comparisons required to decide whether two descriptions refer to the same entity, we distinguish between *pairwise* and *collective* entity resolution. *Pairwise entity resolution* (e.g., [Benjelloun et al., 2009]) compares only two entity descriptions at a time. This comparison depends only on the data contained in these descriptions, and not on the similarity evidence provided by others. *Collective entity resolution* compares a set of related entity descriptions. This comparison heavily relies on similarity evidence provided by neighboring entity descriptions. In a pure collective manner, entity resolution utilizes relationships among descriptions in order to decide about possible matches (e.g., [Bhattacharya and Getoor, 2006, Dong et al., 2005, Kalashnikov and Mehrotra, 2006]). In an iterative manner, entity resolution iterates over the set of current matches and, as match decisions are made, they are used to prompt further match decisions (e.g., [Böhm et al., 2012, Rastogi et al., 2011]).

It is worth noticing that an alternative notion of collectiveness is considered by machine learning techniques, which examine all or part of the entity collection to learn how two descriptions can match. This line of work includes clustering techniques (e.g., [Chaudhuri et al., 2005, McCallum et al., 2000]) and classifiers (e.g., Bayesian networks [Verykios et al., 2003]) and is outside the scope of this lecture. For an exhaustive review on collective entity resolution, readers are referred to [Doan et al., 2012, Dong and Srivastava, 2015, Getoor and Machanavajjhala, 2013, Rastogi et al., 2011].

Furthermore, it is often useful to distinguish between multiple descriptions of the same real-world entities published within or across KBs. KBs with similar descriptions of the same entities are referred to as *dirty*, as opposed to *clean* KBs. In this respect, when an entity resolution task considers as input a dirty entity collection, it is called *dirty entity resolution*, while when it considers two clean, but possibly overlapping entity collections, it is called *clean-clean entity resolution*. Traditionally, the former task targets deduplication in data warehouses [Naumann and Herschel, 2010], while the latter interlinking, eventually by third parties, of entities described by different KBs in the Web of data. When there is a need to combine or merge duplicate entity descriptions to produce a single, possibly more complete representation of that real-world entity, special-purpose data or knowledge fusion functions are used. Since we are focusing more on entity interlinking, this second step is not covered in this lecture. Instead, we refer to [Dong and Naumann, 2009, Dong et al., 2014b] for more details on this topic.

Formal Definition.   To abstract from the syntax of concrete data models (e.g., RDF, relational) used to describe real-world entities, we represent an *entity description* as a set of attribute-value pairs. Formally, let $\mathcal{N}$ be a set of attribute names and $\mathcal{V}$ be a set of values. Let also $\mathcal{E}$ be a collection of entity descriptions and D be the domain of entity descriptions. An entity description $e_i \in \mathcal{E}$ is defined as: $e_i = \{(a_{i_j}, v_{i_j}) | a_{i_j} \in \mathcal{N}, v_{i_j} \in \mathcal{V}\}$. We refer to the set of attributes along with their domains in e_i as the structural type of e_i. We also refer to the set $V_{e_i} = \{v_{i_j} | (a_{i_j}, v_{i_j}) \in e_i\}$ as the values of e_i. For example, the set of attribute-value pairs describing an entity in the Web of data, essentially groups a collection of RDF triples per subject URI (i.e., the entity name).

Entity resolution aims to discover descriptions referring to the same real-world entity, called *matches*. Let us consider a pairwise comparison approach and let M be a match function, determining whether two descriptions $e_i, e_j \in \mathcal{E}$ refer to the same entity. That is, M maps each pair of entity descriptions (e_i, e_j) to *{true, false}*. $M(e_i, e_j) = true$, means that e_i, e_j are matching descriptions; we denote this as $e_i \approx e_j$. $M(e_i, e_j) = false$, indicates that e_i, e_j do not match; we denote this as $e_i \neq e_j$. Formally:

Definition 2.2   Entity resolution.   Let $\mathcal{E} = \{e_1, \ldots, e_m\}$ be a set of entity descriptions and $M : D \times D \to \{true, false\}$ be a boolean match function. An entity resolution of $\mathcal{E}$ is a partitioning $P = \{p_1, \ldots, p_n\}$ of $\mathcal{E}$, such that:

(i) $\forall e_i, e_j \in \mathcal{E} : M(e_i, e_j) = true, \exists p_k \in P : e_i, e_j \in p_k$, and

(ii) $\forall p_k \in P, \forall (e_i, e_j) \in p_k, \ M(e_i, e_j) = true$.

Intuitively, matching entity descriptions are placed in the same partition of P and all the descriptions of the same partition match. Going back to our example of Figure 2.1, entity resolution will ideally place similar descriptions that refer to *Eyes Wide Shut*, *A Clockwork Orange*, *Stanley Kubrick*, and *Manhattan*, respectively, and only those, in four distinct partitions of P.

Strictly speaking, the match function should introduce an *equivalence relation* among entity descriptions[1] as follows [de Melo, 2013]:

(i) $M(e_i, e_i) = true$ (reflexivity),

(ii) $M(e_i, e_j) = M(e_j, e_i)$ (symmetry), and

(iii) $M(e_i, e_j) = true \wedge M(e_j, e_k) = true \Rightarrow M(e_i, e_k) = true$ (transitivity).

In practice, the match function is defined via a similarity function *sim*, measuring how *similar* two entity descriptions are to each other, according to certain criteria. When the similarity between the descriptions e_i, e_j is over a threshold value θ, then $M(e_i, e_j) = true$. Otherwise, $M(e_i, e_j) = false$. Specifically:

$$M(e_i, e_j) = \begin{cases} true, & \text{if } sim(e_i, e_j) \geq \theta \\ false, & \text{else} \end{cases}, \text{ where } e_i, e_j \in \mathcal{E}.$$

In general, specifying the threshold value θ, especially when dealing with somehow similar descriptions, is not trivial. To overcome the problem of setting such a threshold there have been works on machine learning that try to automatically tune this threshold (e.g., [Bilenko and

---

[1]These formal properties stem from a strong notion of identical descriptions for which both *reflexivity* ($e = e$) and *indiscernibility of identicals* ($a = b \to p(a) = p(b)$, for any property p) hold.

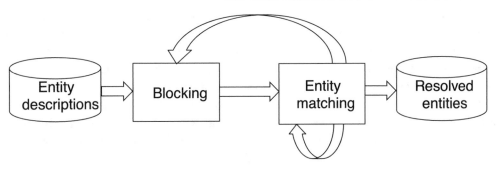

Figure 2.2: Entity resolution process.

Mooney, 2003, Christen, 2008]). Those methods reach their limits in a Web-scale entity resolution, since they have shown significantly worse efficiency and scalability than non-learning methods [Köpcke et al., 2010].

The underlying intuition for this definition is that the higher the similarity of two descriptions, the more likely it is that they match, i.e., the similarity of two descriptions is used as a hint for their matching. Furthermore, given the inherent incompleteness of entity descriptions in the Web of data, two descriptions can only be compared using a part of their attribute-value pairs. This assumption relaxes the strict conditions of the genuine identity relation that should hold between pairs of descriptions and, as a result, the formal properties that a match function should satisfy. Therefore, weakening the strict identity requirements leads to a notion of near-identity or strong similarity [de Melo, 2013]. Highly similar descriptions are more likely to be identified as matches, while it is more difficult to spot somehow similar descriptions, corresponding to matches.

Entity Resolution Process and Challenges.   Figure 2.2 illustrates the general steps involved in an entity resolution process. The core task of entity resolution (Definition 2.2) is to decide whether two descriptions match using an adequate similarity function. For specific domains and relatively small number of KBs, such similarity functions can be easily defined eventually using experts' knowledge. However, the *high semantic and structural heterogeneity* of entity descriptions published in the Web of data (see Chapter 1) make similarity computation a really complex task. In such cross-domain and large-scale entity resolution, even deciding which is the most appropriate piece of descriptions for performing comparisons is an open research issue. For example, do we need to care only for the values of the descriptions, or should we consider any graph structuring of descriptions? What is a reasonable trade-off for assessing similarity between the content-based and structure-based similarity of two descriptions? Moving one step forward, how does schematic information, in terms of employed attribute names and types, affect the degree of similarity (high and somehow) of two descriptions?

Even if we assume that we can answer the above questions, the *very large volume* of entity collections that we need to resolve in the Web of data is prohibitive when examining pairwise all descriptions. In this respect, *blocking* is typically used as a pre-processing step for entity resolution to reduce the number of required comparisons. Specifically, it places similar entity descriptions into blocks, leaving to an entity resolution algorithm the comparisons only between descriptions within the same block. Its goal is to place as many matching descriptions as possible in common blocks, i.e., identify many matches, and only miss as few matches as possible. The underlying assumption is that blocking allows us to disregard comparisons between descriptions that are unlikely to be matches. For further reducing the number of comparisons to be performed by an entity resolution task, blocking techniques can be accompanied by block post-processing steps. Such steps make sense to be used, when blocking results in missing only few matches, and the whole process is faster than exhaustively performing the comparisons between all descriptions. A natural question that arises is what is a good blocking technique for resolving entities in the Web of data. Blocking aims at significantly reducing the number of comparisons, which possibly leads to many missing matches. Overall, it is not straightforward to attain the best trade-off between pruning many comparisons, while retaining the comparisons between matches, since it is not easy to select, or even construct, the appropriate similarity function to use.

To minimize the missed matches, an iterative entity resolution process can exploit in a *pay-as-you-go* fashion any intermediate results of blocking and matching, discovering new candidate description pairs for resolution. Such an iterative process may consider similarity evidence provided by entity descriptions placed into the same block or being structurally related in the original entity graph. We believe that an iterative approach is suitable for coping with the *varying data quality* (e.g., incompleteness) and *loose structuring* (e.g., diverse entity graphs) of entity descriptions in the Web of data.

2.2   SIMILARITY FUNCTIONS

As we have seen in the previous section, the objective of entity resolution is to find the set $\mathcal{M}$ of pairs of descriptions that correspond to the same real-world entities. This knowledge about matching real-world entities is estimated by the set $\mathcal{S}$ of pairs of descriptions that, according to a similarity function, correspond to the same entities. As we can see in Figure 2.3, the choice of the similarity function determines the quality of estimating $\mathcal{M}$ by $\mathcal{S}$. An ideal similarity function will make the two sets coincide, but this is impossible to obtain for all kinds of data [Wang et al., 2011]. Thus, realistic similarity functions typically lead to discover only a fraction of the matches, as well as some non-matches. Their effectiveness is measured in terms of the fraction of the matches identified (i.e., recall, $|\mathcal{M} \cap \mathcal{S}|/|\mathcal{M}|$) and the fraction of the suggested matches that are correct (i.e., precision, $|\mathcal{M} \cap \mathcal{S}|/|\mathcal{S}|$). Hence, in practice, we target at maximizing the intersection of the Venn diagram (i.e., identified matches), while minimizing the difference with the set of matching pairs (i.e., non-matches suggested as matches).

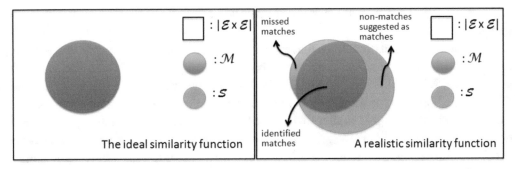

Figure 2.3: Ideal and realistic similarity functions in entity resolution.

Clearly, the quality of estimating $\mathcal{M}$ by $\mathcal{S}$ depends both on the characteristics of the similarity function employed to compare entity descriptions, i.e., which is the piece of information used for comparisons and how, and the underlying formal properties of this similarity function. Such formal properties, widely used in geometric models, are those defining a metric space. Specifically, similarity and distance metrics are defined as follows:

Definition 2.3 Distance Metric.   A distance function $dist : X \times X \to \mathbb{R}$ is a metric, if for any $x, y, z \in X$ for a given set X, it satisfies the following conditions:

1. $dist(x, y) \geq 0$ (non-negativity),

2. $dist(x, y) = dist(y, x)$ (symmetry),

3. $dist(x, y) = 0 \Leftrightarrow x = y$ (identity of indiscernibles), and

4. $dist(x, z) \leq dist(x, y) + dist(y, z)$ (triangle inequality).

Until recently, no formal metric definition for a similarity function had been given. In [Chen et al., 2009], a similarity metric is defined as follows:

Definition 2.4 Similarity Metric.   A similarity function $sim : X \times X \to \mathbb{R}$ is a metric, if for any $x, y, z \in X$ for a given set X, it satisfies the following conditions:

1. $sim(x, x) \geq 0$ (non-negativity),

2. $sim(x, y) = sim(y, x)$ (symmetry),

3. $sim(x, x) \geq sim(x, y)$,

4. $sim(x, x) = sim(y, y) = sim(x, y) \Leftrightarrow x = y$, and

5. $sim(x, y) + sim(y, z) \leq sim(x, z) + sim(y, y)$ (triangle inequality).

[Chen et al., 2009] proves that any similarity metric can be transformed to a distance metric and vice versa.

Defining similarity functions that satisfy the formal properties of metric spaces is, in practice, too restrictive for non-geometric models [Jacobs et al., 2000, Mu and Yan, 2010, Santini and Jain, 1999, Skopal, 2006, Tversky, 1977]. For example, the triangle inequality assumes that the notion of similarity is transitive. A well-known counter-example is that a man is similar to a centaur and the centaur is similar to a horse; however, the man is completely dissimilar to the horse [Mu and Yan, 2010, Skopal, 2006]. The suggestion that the assessment of similarity between entity descriptions may be better described as a comparison of their attribute-value pairs [Tversky, 1977] is supported by experimental works [Jacobs et al., 2000], reporting that in specific domains, non-metrics show better results than metrics. Even so, as [de Melo, 2013] claims, there is no general way of determining which attributes should count as salient in determining matching entity descriptions. As studied in [Hogan et al., 2010], a pair of descriptions is more likely to be matching if they share several common attribute-value pairs, while certain attributes are more appropriate to determine matches and certain values of these attributes are more discriminant than others.

Overall, the formal properties of metric spaces appear to be applicable to duplicates, i.e., identical, or highly similar descriptions of the same real-world entity. However, matching entity descriptions in the Web of data, due to their high heterogeneity, are commonly only somehow similar, making the satisfaction of the metrics properties too rigid. Motivated by this observation, entity resolution and blocking algorithms disregard similarity and distance metrics, in favor of non-metrics [Araújo et al., 2012, Böhm et al., 2012, Papadakis et al., 2013, Zhang et al., 2012b].

In the remainder of this chapter, we will survey the main similarity functions that have been employed to resolve entity descriptions in the Web of data. They compare descriptions either exclusively on their content (i.e., attribute values), or additionally on their structural relations with other descriptions. Regarding the content-based similarities, *character-based* similarity functions, taking as input a pair of strings, are very useful when comparing the values of a fixed set of attributes, as in traditional data warehouses, where the only data variations occur in the values of a given schema. However, such an assumption cannot be made for entity descriptions in the Web of data, where descriptions are loosely structured, as analyzed in Chapter 1. Hence, to overcome this problem, many works in the Web of data use more flexible *token-based* similarity functions, operating on all the values of a description, split as a set of tokens or *n*-grams. From a different point of view, *information-theoretic* similarity functions exploit probability distributions, extracted from data statistics. Regarding the similarity functions that also consider the structural relations of descriptions, we divide them into *tree-based*, mainly applied to the relational star/snowflake schema or XML data, and *graph-based*, applied to RDF data. Motivated by the fact that, in practice, only pairs of descriptions with a high enough similarity are of interest, we will present popular *approximations of similarity functions*, addressing the scalability hurdles that arise when comparing a large number of descriptions.

2.2.1   CONTENT-BASED SIMILARITY FUNCTIONS

Character-based similarity functions   Character-based functions manage to account for typographical errors (character swaps, typos, etc.). In general, these functions allow edit operations in order to transform one string into another, for example, by inserting, deleting, or substituting characters.

The *Levenshtein distance* [Levenshtein, 1966], also known as *edit distance*, of two strings a and b, $Levenshtein(a, b)$ is given by $lev_{a,b}(|a|, |b|)$, where:

$$lev_{a,b}(i, j) = \begin{cases} \max(i, j), \text{ if } \min(i, j) = 0 \\ \min \begin{cases} lev_{a,b}(i - 1, j) + 1 \\ lev_{a,b}(i, j - 1) + 1 \\ lev_{a,b}(i - 1, j - 1) + 1_{(a_i \neq b_i)} \end{cases} , \text{otherwise.} \end{cases}$$

$1_{(a_i \neq b_j)}$ is the indicator function equal to 0, when $a_i = b_j$, and equal to 1, otherwise. Intuitively, the *Levenshtein* of two strings is the minimum number of single-character insertions, deletions or substitutions required to change one string into the other. The normalized version of this similarity, $LEV_{sim}(a, b)$, is used in [Böhm et al., 2012, Zhang et al., 2012b] to measure the similarity of two string labels a, b.

$$LEV_{sim}(a, b) = 1 - \frac{Levenshtein(a, b)}{\max(|a|, |b|)}.$$

The *Jaro similarity* [Jaro, 1989] of two strings a, b is defined with respect to their common characters. Once such characters are identified, we count the number of their transpositions in a, b. That is,

$$Jaro(a, b) = \frac{1}{3}\left(\frac{|c|}{|a|} + \frac{|c|}{|b|} + \frac{|c| - 0.5t}{|c|}\right),$$

where $|c|$ is the number of common characters in a and b; two characters are considered to be common, if they are the same and their positions within the two strings do not differ by more than $max(|a|, |b|)/2$. t is the number of transpositions occurring when the i-th common character of a is not the same as the i-th common character of b. In general, Jaro performs well for strings with few variations. In a similar manner, the *Jaro-Winkler similarity* [Winkler, 1999] of a, b is computed taking into account their longest common prefix.

The *Levenshtein distance* satisfies the formal properties of a metric, but *Jaro* and *Jaro-Winkler similarity* functions are not symmetric, and consequently, cannot be considered as metrics [Euzenat and Shvaiko, 2013].

Token-based similarity functions   Token-based functions take as input the set of tokens of two descriptions, or alternatively, the set of n-grams of these descriptions, i.e., substrings of length n. Interestingly, such functions are not sensitive to the order of tokens or n-grams that are compared, which means that comparing *Auguste Bartholdi* and *Bartholdi Auguste* leads to the maximum similarity score, since both sets contain exactly the same tokens.

A widely used token-based function is the *Jaccard similarity* that compares two sets A, B as follows:

$$Jaccard(A, B) = \frac{|A \cap B|}{|A \cup B|}.$$

Intuitively, Jaccard similarity expresses the number of tokens two sets have in common divided by the total number of unique tokens. Jaccard similarity is used as well in [Papadakis et al., 2013] for computing similarities between attribute names, based on the trigrams in their values.

The *dice similarity* of A, B is defined as:

$$dice(A, B) = \frac{2|A \cap B|}{|A| + |B|}.$$

Similar to Jaccard, the dice similarity is equal to the number of tokens or n-grams in common to both sets, relative to the average size of the total number of tokens present. Intuitively, the more the common tokens, the higher the similarity of Jaccard compared to dice.

The *overlap similarity* of two sets A, B is defined as:

$$overlap(A, B) = \frac{|A \cap B|}{\min(|A|, |B|)}.$$

Abstractly, overlap similarity counts the number of tokens two sets have in common, divided by the number of tokens in the smaller set. Overlap is a looser function than Jaccard and dice, disregarding the difference in the size of the two sets. Readers are referred to Section 4.2.4 of [Augsten and Böhlen, 2013] for the formal definition of the equivalence between Jaccard, dice, and overlap similarity.

The *cosine similarity* additionally exploits some statistics. It compares vectors, weighted with the TF-IDF model [Dhillon and Modha, 2001], representing tokens. Intuitively, the idea is that two entities are more similar if they share a token that is rare in the collection. Specifically, *cosine* between two vectors $\vec{A}$, $\vec{B}$ is defined as:

$$cosine(\vec{A}, \vec{B}) = \frac{\vec{A} \cdot \vec{B}}{||\vec{A}|| \, ||\vec{B}||}.$$

A variation of cosine similarity uses soft TF-IDF for constructing the vectors, in which model tokens are regarded as equal, when their edit distance is small and not necessarily zero. Cosine similarity with TF-IDF weights has been used in [Papadakis et al., 2013] to compute similarities between attribute names with respect to the vectors of their values. If A and B represent sets of tokens, then cosine similarity can be also written as:

$$cosine(A, B) = \frac{|A \cap B|}{\sqrt{|A||B|}}.$$

All the token-based similarity functions are of the form:

$$F(A, B) = \frac{\psi_1(|A \cap B|)}{\psi_2(|A|, |B|, |A| \cup |B|)},$$

where ψ_1 is a strictly increasing function and ψ_2 is an increasing function of three variables [Egghe and Michel, 2003]. We can classify a token-based similarity function F as *strong* or *weak*, based on the following set of properties [Egghe and Michel, 2003]:

($P1$) $0 \leq F(A, B) \leq 1$.

($P2$) $F(A, B) = 1 \Leftrightarrow A = B$.

($P3$) $F(A, B) = 0 \Leftrightarrow A \cap B = \emptyset$.

($P4$) If the denominator of F is constant then F is strictly increasing with $|A \cap B|$.

($P1$) is similar to the *non-negativity* of metric spaces, additionally setting the value 1, as the upper threshold of similarity. ($P2$) is semantically close to the *identity of indiscernibles*, stating that two descriptions have an ideal similarity if and only if they are identical. ($P3$) and ($P4$) determine the importance of the intersection for token-based similarity functions. ($P3$) states that an empty-set intersection means zero similarity, while ($P4$) requires a greater intersection between two descriptions to imply a greater similarity between them, if the denominator is unchanged. A *strong* similarity function is one that satisfies all these four properties. For example, Jaccard, dice and cosine similarity functions are strong. On the other hand, if ($P2$) is replaced by the weaker:

($P2'$) $F(A, B) = 1 \Leftrightarrow A \subset B \ or \ B \subset A$,

meaning that two descriptions can have an ideal similarity, even if they are not identical, then a *weak* similarity function is one that satisfies ($P1$), ($P2'$), ($P3$), and ($P4$), like the overlap similarity function. Note that *symmetry* and *triangle inequality* are not required for token-based similarity functions to be labeled as strong or weak. Note also that only Jaccard can be considered as a metric [Clarkson, 2006, Jacox and Samet, 2008], since the rest of the token-based similarity functions do not satisfy the triangle inequality [Charikar, 2002].

Information-theoretic similarity functions   From a different point of view, similarity functions used in information theory can quantify the statistical relationship between two descriptions, such as their interdependency. To accomplish that, these measures exploit statistics extracted from the attributes or values of the entity collections, e.g., their co-occurrences.

The *mutual information* measures the mutual dependence of two random variables A, B. That is:

$$MI(A; B) = \sum_{b \in B} \sum_{a \in A} p(a, b) \ ln \left(\frac{p(a, b)}{p(a) p(b)} \right),$$

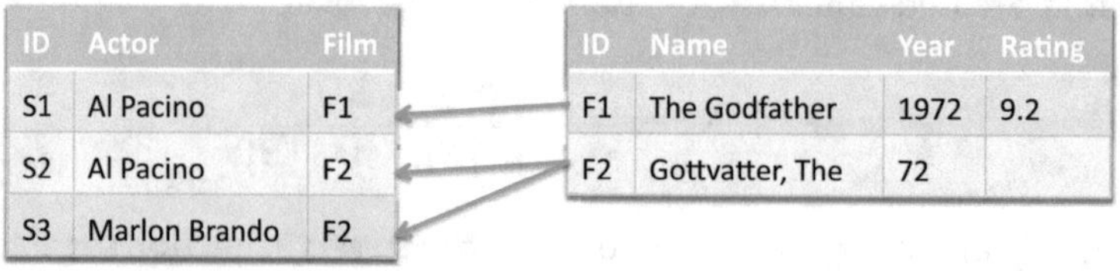

ID	Actor	Film
S1	Al Pacino	F1
S2	Al Pacino	F2
S3	Marlon Brando	F2

ID	Name	Year	Rating
F1	The Godfather	1972	9.2
F2	Gottvatter, The	72	

Figure 2.4: An example of a relational star schema.

where $p(a, b)$ is the joint probability distribution of A and B and $p(a)$ and $p(b)$ are the marginal probability distributions of A and B, respectively. Intuitively, mutual information measures the information that A and B share: it measures how much knowing one of these variables reduces uncertainty about the other. For example, if A and B are independent, then knowing A does not give any information about B and vice versa, so their mutual information is 0. At the other extreme, if A is a deterministic function of B and B is a deterministic function of A, then all information conveyed by A is shared with B, which means that knowing A determines the value of B and vice versa. In the context of Web tables, [Cafarella et al., 2008] uses the *pointwise mutual information*, *PMI*, to measure the coherency score of two schemas, i.e., how well attributes a, b of a schema fit together, defined as:

$$PMI(a, b) = \ln \left(\frac{p(a, b)}{p(a)p(b)} \right).$$

2.2.2 RELATIONAL SIMILARITY FUNCTIONS

All functions discussed so far consider only the descriptions to be compared. Yet, additional information regarding the relationships of the descriptions with other descriptions can be used as well. We say that two descriptions are *neighbors*, if they are somehow linked, i.e., by using a foreign key in relational databases, a parent-child relationship in XML data, or a triple linking the URIs of two descriptions in RDF. Here, we divide relational similarity functions into *tree-based*, most notably applied to relational star/snowflake schema or XML data, and *graph-based*, applied to RDF data. Relational similarity functions are typically expressed as a linear combination of a value similarity and a neighborhood similarity of descriptions.

Tree-based functions: When considering tree or hierarchically structured data, new challenges arise due to the relationships between the descriptions. In our discussion, without loss of generality, we resort to a relational star schema (see for example, the toy relational tables of Figure 2.4).

The DELPHI containment metric [Ananthakrishna et al., 2002] is a specialized similarity function that can be applied to data of such type. Given two entity descriptions, in the form of tuples, DELPHI takes into account both the similarity of their attribute values, *tcm*, and the similarity of their children sets reached by following foreign keys, *fkcm*. For computing *tcm*, we first divide tuples into token sets TS and then compute the Levenshtein distance of the token sets. Tokens are weighted using the IDF model. Overall, *tcm* measures which fraction of one tuple T is covered by the other tuple T':

$$tcm(T, T') = \frac{\sum idf(TS(T) \cap TS(T'))}{\sum idf(TS(T))}.$$

In turn, *fkcm* measures at what extent the children set CS of a tuple T is covered by the children set of a tuple T', where CS of T includes all tuples referencing T from other relations by means of a foreign key:

$$fkcm(T, T') = \frac{\sum |CS(T) \cap CS(T')|}{\sum |CS(T)|}.$$

DELPHI combines *tcm* and *fkcm*, by using a classification function, as follows:

$$pos(idf(TS) \times pos(tcm(T, T') - s_1) + idf(CS) \times pos(fkcm(T, T') - s_2)),$$

where s_1, s_2 are threshold values for *tcm*, *fkcm*, respectively, and $pos(x) = 1$, if $x > 0$, or -1, otherwise. If the result equals 1, then the tuples match, otherwise they do not match.

Example 2.5   Given the tuples $F1$ and $F2$ of Figure 2.4, $TS(F1) = \{The, Godfather, 1972, 9.2\}$ and $TS(F2) = \{Gottvatter, The, 72\}$. For simplification, assume that all tokens have equal weight, and the distance of the token sets is negligible, i.e., consider that Godfather = Gottvatter and 1972 = 72. Then, $tcm(F1, F2) = 3/4$ and $tcm(F2, F1) = 1$. Similarly, $fkcm(F1, F2) = 1$ and $fkcm(F2, F1) = 1/2$.

$F1$ and $F2$ match for $s_1 = s_2 = 0.5$ and weights = 1, since $pos(pos(3/4 - 0.5) + pos(1 - 0.5)) = 1$.

[Weis and Naumann, 2005] adapts DELPHI and introduces a symmetric function, unlike DELPHI, that manages to work not only for a single branch of hierarchy, but also for cases in which one table may have several children tables.

Graph-based functions: Compared to hierarchical data, graph-structured data poses further challenges when calculating similarities between entity descriptions. Specifically, the procedure of computing similarities becomes more complicated, since in addition to the existing relationships between descriptions, cycles between them occur as well. Interestingly, the functions of this category provide a means for handling the structural heterogeneity of data, toward resolving entities that appear in the Web.

A graph-based similarity function can be simply built on the number of common neighbors that two descriptions share. [Bhattacharya and Getoor, 2007] evaluate various such functions, with the simplest one being the number of common neighbors that two descriptions have, normalized by an adequately large constant number, such that all the scores in the collection are less than 1. A better normalization is offered by employing the Jaccard similarity for the sets of neighbors of two descriptions. Alternatively, the Adar similarity [Adamic and Adar, 2003] of two descriptions e_i, e_j is defined as:

$$Adar(e_i, e_j) = \frac{\sum_{e \in N(e_i) \cap N(e_j)} u(e)}{\sum_{e \in N(e_i) \cup N(e_j)} u(e)},$$

where $N(e)$ is the set of neighbors of a description e, and $u(e) = \frac{1}{log(|N(e)|)}$ counts the uniqueness of e, which is inversely proportional to its number of neighbors.

Recently, LINDA [Böhm et al., 2012] proposes a similarity function for comparing two entity descriptions taking into account: (i) the similarity of their values, and (ii) the similarity of their neighbors. In particular, assume an entity graph G, i.e., an RDF graph whose nodes (entity descriptions) are only URIs, thus excluding literals and blank nodes. An assignment matrix X is a square binary $\mathcal{E} \times \mathcal{E}$ matrix, representing the knowledge for the already identified matches between descriptions. Given G, X, the similarity, $LINDA_{sim}$, of two descriptions e_i, e_j is defined as:

$$LINDA_{sim}(e_i, e_j, G, X) = L_{prior}(V(e_i), V(e_j)) + \alpha \, L_{context}(C(e_i), C(e_j), G, X) - \theta,$$

where L_{prior} is a token-based similarity function on the literal values of e_i, e_j, $L_{context}$ is a similarity function on the neighbors of e_i and e_j in G, α is an empirically-tuned weight of the $L_{context}$ similarity, and θ ensures that the scores are renormalized to values around 0. $LINDA_{sim}$ is not a normalized measure and it serves as a means of ranking pairs of descriptions, based on the evidence that they are matching. The more common tokens and common neighbors that two descriptions have, the more likely they are to match.

More specifically, L_{prior} is applied to the sets of literal values of the two descriptions. It is an extension of the overlap similarity, that takes the set size difference into account. Unlike Jaccard similarity, L_{prior}'s normalization is biased toward the size of the smaller set and, hence, better accounts for data heterogeneity. The reason for this bias is that, in the Web of data, the sizes of the two sets can vary significantly, e.g., due to incomplete information provided for an entity description in a KB, as explained in Chapter 1. Given two sets A, B:

$$L_{prior}(A, B) = \frac{|A \cap B|}{min(|A|, |B|) + ln(|(|A| - |B|)| + 1)}.$$

The $L_{context}$ similarity of the descriptions e_i, e_j is:

$$L_{context}(C(e_i), C(e_j), G, X) = \sum_{\substack{(r_{e_i}, n_{e_i}, w_{e_i}) \\ \in C(e_i)}} \max_{\substack{(r_{e_j}, n_{e_j}, w_{e_j}) \\ \in C(e_j)}} x_{n_{e_i}, n_{e_j}} \cdot w_{e_i} \cdot LEV_{ssim}(r_{e_i}, r_{e_j}).$$

The context $C(e_i)$ of e_i in G is a set of $(r_{e_i}, n_{e_i}, w_{e_i})$ tuples, where r_{e_i} is a property, i.e., an edge in the entity graph, connecting e_i to its neighbor n_{e_i} and w_{e_i} is the weight of this tuple, based on

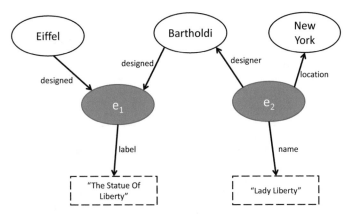

Figure 2.5: An example RDF graph, used to evaluate $LINDA_{sim}$.

how discriminative this (r_{e_i}, n_{e_i}) pair is. x_{e_i,e_j} represents the position (e_i, e_j) of the assignment matrix X. Intuitively, $L_{context}$ finds matching pairs of context tuples and sums up their similarity values. Essentially, it counts the number of common or matching neighbors of two descriptions, which are linked to them in a similar way, i.e., using a (character-based) similar property.

Example 2.6   Consider the RDF graph G of Figure 2.5. In order to compute the similarity between e_1 and e_2, we first compute L_{prior}, which only takes literal values into account. The literal values of e_1 are the set $\{The, Statue, Of, Liberty\}$, and of e_2 the set $\{Lady, Liberty\}$, so, $L_{prior}(V(e_1), V(e_2)) = \frac{1}{2+ln(3)} = 0.37$.

Moving on to $L_{context}(C(e_i), C(e_j), G, X)$, we sum up the maximum similarities of each of the two neighbors of e_1 (Eiffel and Bartholdi) to the neighbors of e_2 (New York and Bartholdi). X is initialized as a diagonal matrix, meaning that, at first, we only know that a description matches only with itself. So, initially, $x_{e_i,e_j} = 1$, only when $e_i = e_j$. Thus, we only consider the contextual similarity of e_1 and e_2 with respect to their common neighbor (Bartholdi) at first.

$$L_{context}(C(e_i), C(e_j), G, X) = x_{Bartholdi,Bartholdi} \cdot w_{"designed"} \cdot LEV_{sim}("designed," "designer") =$$

$$= 1 \cdot \frac{1}{log(freq("designed," e_1))} \cdot (1 - \frac{1}{8}) = \frac{1}{log(2)} \cdot \frac{7}{8} = 0.875.$$

Finally, the overall $LINDA_{sim}$ of e_i and e_j can be calculated as: $LINDA_{sim}(e_i, e_j, G, X) = 0.37 + 2 \cdot 0.875 - 0.7 = 1.42$, given that the default, empirically suggested values for α and θ are 2 and 0.7, respectively. w is computed, here, as $1/log(freq(r, n))$, where $freq(r, n)$ is the total number of occurrences of property r with entity n.

[Zhang et al., 2012b] presents a similarity function used for comparing descriptions in an RDF graph, taking as well into consideration the similarity of their neighbors in the graph. To compute the similarity of two descriptions, this function decomposes each description to its

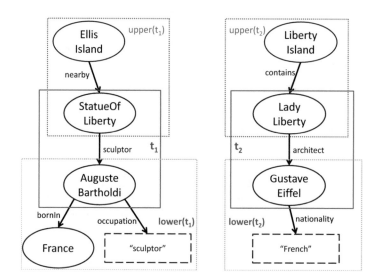

Figure 2.6: An example RDF graph, used to evaluate sim_{ov}.

set of RDF triples and computes pairwise similarities between the triples of each description. Specifically, assume the triple t_i of a description e_k. The upper set of t_i is composed of all the triples that have the subject of t_i as their object, i.e., all the triples of the in-neighbors of e_k that link them to e_k. The lower set of t_i is composed of all the triples that have the object of t_i as their subject, i.e., all the out-neighbors of e_k. Then, the upper set similarity sim_{upper} (and lower set similarity sim_{lower}, respectively) between two triples t_i, t_j is defined as:

$$sim_{upper}(t_i, t_j) = \frac{\sum\limits_{x \in upper(t_j)} \max\limits_{y \in upper(t_i)} sim_{triple}(x, y)}{|upper(t_j)|},$$

where $sim_{triple}(t_i, t_j)$ is a weighted mean of LEV_{sim} between the labels of the subjects, predicates, and objects of t_i and t_j. Implicitly, by comparing the triples predicates, sim_{upper} utilizes structural information about the triples. Then, the structural similarity of (t_i, t_j) is defined as a weighted summation of their upper and lower sets similarities: $sim_{ul}(t_i, t_j) = \beta \, sim_{upper}(t_i, t_j) + (1 - \beta) \, sim_{lower}(t_i, t_j)$. Finally, the overall similarity sim_{ov} of t_i and t_j is given by: $sim_{ov}(t_i, t_j) = \alpha \, sim_{triple}(t_i, t_j) + (1 - \alpha) \, sim_{ul}(t_i, t_j)$, where α is the weight given to the similarity of the labels, over the similarity of neighbors. To capture the similarity between two entity descriptions, we can then aggregate the pairwise similarities of their triples.

Example 2.7   Consider the example of Figure 2.6, in which we want to compare e_i, consisting of the triple t_1 = (StatueOfLiberty, sculptor, AugusteBartholdi) to e_j, consisting of the triple t_2

= (LadyLiberty, architect, GustaveEiffel). To compute the overall similarity of e_i and e_j, we first compute the value of $sim_{triple}(t_1, t_2)$, assigning equal weights to subject, predicate, and object. Thus:

$sim_{triple}(t_1, t_2) = 1/3\ LEV_{sim}(\text{"StatueOfLiberty,"\ "LadyLiberty"}) + 1/3\ LEV_{sim}(\text{"sculptor,"}$
$\text{"architect"}) + 1/3\ LEV_{sim}(\text{"AugusteBartholdi,"\ "GustaveEiffel"}) = 1/3\ (1 - \frac{7}{15}) + 1/3\ (1 - \frac{8}{9}) + 1/3\ (1 - \frac{13}{16}) = 0.707.$

Then, we compute the similarity of the upper sets of t_1 and $t2$, as: $sim_{upper}(t_1, t_2) =$
$= sim_{triple}((\text{EllisIsland, nearby, StatueOfLiberty}), (\text{LibertyIsland, contains, LadyLiberty})) =$
$= 1/3\ LEV_{sim}(\text{"EllisIsland,"\ "LibertyIsland"}) + 1/3\ LEV_{sim}(\text{"nearby,"\ "contains"}) +$
$1/3\ LEV_{sim}(\text{"StatueOfLiberty,"\ "LadyLiberty"}) = 0.889.$ Similarly, $sim_{lower}(t_1, t_2) =$
$= \max(sim_{triple}((\text{GustaveEiffel, nationality, "French"}), (\text{AugusteBartholdi, bornIn, France})),$
$sim_{triple}((\text{GustaveEiffel, nationality, "French"}), (\text{AugusteBartholdi, occupation, "sculptor"}))) = 0.5.$ So, assuming equal weights again, $sim_{ul}(t_1, t_2) = (0.889 + 0.5)/2 = 0.695.$ Finally, the overall similarity of e_i and e_j is: $sim(e_i, e_j) = sim_{ov}(t_1, t_2) = (0.707 + 0.695)/2 = 0.701.$

2.2.3   APPROXIMATIONS OF SIMILARITY FUNCTIONS

When resolving very large collections of entity descriptions in the Web of Data, two big data processing issues are arising: (a) how can we shorten the representation of entity descriptions in main-memory and (b) how can we avoid systematically comparing all pairs of descriptions for similarity? To address these issues, approximation techniques have been used both for converting large sets of tokens, extracted by entity descriptions, to short signatures, as well as for locating pairs of signatures that are likely to be similar [Duan et al., 2012, Kim and Lee, 2010]. These techniques essentially adapt pioneering work for detecting near-duplicate documents on the Web that take into account the ordering of words [Broder, 1997, Broder et al., 1998] (e.g., mirror websites, similar news articles from different press agencies, etc.). They rely on appropriate hash functions that preserve as much as possible the original similarity of documents (e.g., Jaccard). Due to their approximate nature, these techniques entail false negatives and even false positives and require a thorough tuning of their configuration parameters to achieve a reasonable trade-off for a particular collection of entity descriptions. In the sequel, we will briefly present the core ideas of these approximation techniques for resolving entities (for more details, readers are referred to [Rajaraman and Ullman, 2011]).

Minhashing   Minhashing is a technique tackling the inherent high-dimensionality of textual comparisons. It has been originally proposed for computing the similarity of documents, represented as small signatures rather than large sequences of k characters (called *k-shingles*). Minhashing can be naturally applied for comparing two entity descriptions based on the set of tokens they contain in their attribute values. To find subsets that have a significant intersection, a Boolean representation of descriptions, called *characteristic matrix* is employed. The columns of the matrix correspond to the descriptions that need to be compared, while the rows to their tokens, from a universal set (extracted from all descriptions). If a description in column c contains a token in

token/description	e1	e2	e3
the	1	0	1
statue	1	1	0
of	1	0	0
liberty	1	1	0
lady	0	1	0
eiffel	0	0	1
tower	0	0	1

	h1	h2	h3
	1	5	6
	2	6	5
	3	3	4
	4	7	2
	5	1	7
	6	4	1
	7	2	3

	e1	e2	e3
h1	1	2	1
h2	3	1	2
h3	2	2	1

(a) (b) (c)

Figure 2.7: An example characteristic matrix (a), three random permutations $h1, h2, h3$ of the rows of this matrix (b), and the resulting minhash signature matrix (c).

row r, then the value of (r, c) is 1; otherwise it is 0. In real application settings, the characteristic matrix is essentially *sparse*. Only a small subset of the tokens from the universal set appears in individual entity descriptions, while two descriptions contain at best only partially overlapping sets of tokens. Minhashing is a hashing technique allowing us to reduce the number of rows (i.e., tokens) that need to be compared between two descriptions. The minhash value of any column is the index of the first row in which the column has a 1, when rows are randomly permuted, i.e., random row permutations play the role of hash functions. By applying n random permutations to the rows of the matrix, a column e_i will be represented by its minhash *signature*, namely $(h_1(e_i), h_2(e_i), \ldots, h_n(e_i))$ where $h_1, h_2, \ldots, h_n$ are the employed row permutations. As we will see in the next section, Locality-Sensitive Hashing exploits the so-constructed minhash signatures to reduce the number of columns (i.e., descriptions) that actually need to be compared for similarity. The underlying intuition is that two descriptions, agreeing on a sufficient number of minhash values, are likely to be similar and are, thus, nominated as candidate pairs for comparison. To overcome the cost of computing row permutations, random hash functions can be used for mapping row numbers to as many buckets as there are rows. Hence, ordering under h_i will give a random row permutation.

Example 2.8   Consider the entity descriptions $e1$ = {the, statue, of, liberty}, $e2$ = {lady, liberty, statue} and $e3$ = {the, eiffel, tower}. Their characteristic matrix is depicted in Figure 2.7 (a), while the signature matrix, produced by minhash for the three random permutations of rows of Figure 2.7 (b), is depicted in Figure 2.7 (c). That is, the minhash of $e1$ for the order specified by $h2$, is 'of', i.e., $h2(e1) = 3$, the minhash of $e2$ is 'lady', i.e., $h2(e2) = 1$, and the minhash of $e3$ is 'tower', i.e., $h2(e3) = 2$. The minhash signature $(1, 3, 2)$ of $e1$ is given by the minhashes of $e1$, i.e., the column corresponding to $e1$.

Jaccard approximation via minhashing   An interesting property of minhashing is that it effectively approximates the Jaccard similarity of two sets. More precisely, the probability that the minhash function for a random permutation of rows produces the same value for two sets is equal to the Jaccard similarity of the sets. The key idea is that for any two sets e_i and e_j, the ratio of rows having 1 as minhash value in both columns (denoted as type A) with the rows having 1 in one of the columns and 0 in the other (denoted as type B) determines both $sim(e_i, e_j)$ and the probability that $h(e_i) = h(e_j)$. Let a be the number of type A rows and b be the number of type B rows; then, $sim(e_i, e_j) = a/(a + b)$. This is because a is the size of $e_i \cap e_j$ and $a + b$ is the size of $e_i \cup e_j$. Now, consider the probability that $h(e_i) = h(e_j)$, given a random permutation of rows. If we traverse the matrix from the top, the probability that we will see a type A row before a type B row is $a/(a + b)$ (ignoring rows with 0 in both columns). But, if the first row from the top is of type A, then $h(e_i) = h(e_j)$. If the first row is of type B, then the description with a 1 gets that row as its minhash value. However, the description with a 0 in that row surely gets some row further down the permuted list. So, we know $h(e_i) \neq h(e_j)$, if we first see a type B row. Thus, overall, the probability that $h(e_i) = h(e_j)$ is $a/(a + b)$, which is also the Jaccard similarity of e_i and e_j. So, we can estimate the Jaccard similarity of two descriptions by computing the ratio of the number of the same minhash values to the number of the minhash functions.

Example 2.9   Using the minhash signature matrix of Figure 2.7 (c), we can estimate the Jaccard similarities of $e1$, $e2$, $e3$. Specifically, from the columns of $e1$ and $e2$, we can assume that $sim(e1, e2) = 1/3$. However, from a closer look at Figure 2.7 (a), the actual Jaccard similarity of $e1$ and $e2$ is $2/5$; remember that minhash offers only an approximation of the Jaccard similarity. By considering more permutations of the rows, i.e., more hash functions, we achieve a better approximation of the actual Jaccard similarity value. If we had only considered $h1$ and $h2$, then we would have estimated the similarity of $e1$ to $e2$ as 0. By adding an extra hash function $h3$, we refine our estimation as $1/3$, which is closer to the actual similarity value. As an additional example, consider that based on the signatures of $e1$ and $e3$, we estimate their Jaccard similarity as $1/3$, while their true similarity is $1/6$. If we had only used $h1$ and $h2$, then we would have estimated their similarity as $1/2$.

Locality-Sensitive Hashing   While the minhash signatures of all columns may fit in main memory, comparing the signatures of all pairs of columns is still very costly (i.e., quadratic in the number of columns). Locality-Sensitive Hashing (LSH) relies on an adequate function that tells us whether or not x and y constitute a candidate pair: a pair of descriptions whose similarity should be evaluated. The key idea is to hash descriptions multiple times, using a family of hash functions, in such a way that similar descriptions are more likely to be placed into the same bucket than dissimilar ones. Any two descriptions that hash at least once into the same bucket, for any of the employed hash functions, are considered to be a candidate pair. The underlying assumption is that there are enough buckets that columns are unlikely to hash to the same bucket, unless they are identical with respect to a particular hash function. Those that do not are false nega-

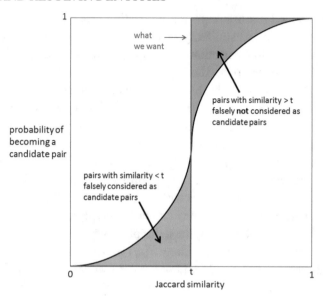

Figure 2.8: The S-curve.

tives; hopefully, these will be only a small fraction of the truly similar pairs. At the same time, LSH targets at not placing most of the descriptions of the dissimilar pairs to the same bucket, and therefore these pairs will never be checked. The dissimilar pairs that are placed to the same bucket constitute false positives, and hopefully there will be only a small number of such pairs.

When the matrix of minhash signatures is available, we can hash its columns multiple times, each time on a different column partition. According to this strategy, we partition the signature matrix into b bands, each one with r rows. For each band, we use a hash function taking vectors of r integers, namely the part of the column within that band, and hash them to a big number of buckets. For each band, we utilize a different bucket, so that columns with the same vector in different bands will not be placed into the same bucket. This way, the similar columns are much more likely to become candidate pairs than dissimilar ones. The probability that two descriptions have the same hash value (of r rows) for the same band, follows the *S-curve* of Figure 2.8. Ideally, for a given Jaccard similarity threshold t, the probability that two descriptions, whose similarity is above t, have at least one band identical should be 1, while in the opposite case, it should be 0. In practice, the step-function is approximated by $(1/b)^{1/r}$. The blue area in Figure 2.8 represents false negatives (i.e., pairs of descriptions above the similarity threshold that do not share a band), while the green area false positives. To achieve a better trade-off according to the data characteristics of the resolved collection of entities, generalizations of the simple LSH strategy have been proposed using families of hash functions [Rajaraman and Ullman, 2011].

2.3   DISCUSSION

In this chapter, we have presented two main research questions when resolving entities in the Web of data:

(a) How can we effectively compare highly diverse descriptions of entities exhibiting different structuring?

(b) How can we efficiently compute the similarity of a very large number of entity descriptions?

Regarding the first research question, we believe that for assessing somehow-similarity of entity descriptions, it is not sufficient to compare the descriptions based only on their content (i.e., their attribute values). We also need to consider contextual information provided, for instance, by their neighbor descriptions in the entity graph. The definition of an adequate weighting scheme for content- and structure-based similarity of entity descriptions remains open and clearly depends on the data characteristics of the resolved entity collection. A pragmatic approach could consider different aspects of similarity among descriptions at different processing steps (see Figure 2.2). For example, content similarity could be useful for bootstrapping the ER process (i.e., blocking), while similarity of their structural neighbors could provide valuable evidence for matching descriptions (e.g., with low content similarity) in a pay-as-you-go way. An orthogonal issue is how schematic discrepancy in terms of employed attribute names and link types affect the content or structural similarity of descriptions. For simple cases, where there is cross-source schema agreement (e.g., by adopting schema.org vocabularies), domain-specific matching rules can be used or learned [Isele and Bizer, 2012]. For other cases, where there is less ontological agreement and ontological alignment is poor, one needs to assess the discriminating ability of different attributes and/or their values and rely on different similarity functions depending on the involved data type if known (e.g., strings vs. numbers vs. dates). Clearly, the higher degree of semantic and structural heterogeneity is exhibited by entity descriptions the more complex becomes the matching task. A promising area of research in this respect is cross-domain similarity search and mining [Dong, 2012, Shrivastava et al., 2011, Zhen et al., 2015], aiming to exploit similarity of objects described by different modalities (i.e., text, image) and contexts (i.e., facets) and support research by analogy. Such techniques could be also beneficial for matching highly heterogeneous entity descriptions and thus support ER at the Web scale.

Regarding the second research question, the main way to reduce the quadratic number of comparisons between descriptions required by entity resolution is by performing a preliminary pruning on the candidate pairs of descriptions. This pre-processing step could be based either on LSH or on blocking, presented in the next chapter. Both require investigating the trade-off between the number of discarded non-matches and the number of missed matches between entity descriptions. However, tuning LSH to achieve a reasonable trade-off (see the S-curve of Figure 2.8) assumes an a priori knowledge of a minimum similarity threshold between entity description pairs, above which, such pairs are considered as candidate matches. As we will see in Chapter 5, often, matching descriptions do not share many common tokens and thus, have very low similarity when computed only on the values of their attributes. Those matches would not be placed in the same bucket by any threshold-based approach, e.g., LSH, and thus, they would not be considered as candidate pairs. Effectively choosing a minimum similarity threshold also depends on the KBs. For example, when seeking matches between two central KBs, a high similarity threshold can be used, since such KBs usually have more similar values. Using a lower threshold in central KBs would result in many false candidate pairs. Accordingly, using a high similarity threshold in peripheral KBs, in which descriptions have lower similarity values, would yield many missed matches (as threshold t goes to 0, the blue area in Figure 2.8 becomes significantly bigger than the green one). Consequently, applying LSH across the domains of the Web is an open research problem, due to the difficulty in knowing or tuning a similarity threshold that can be generalized to identify matches across several domains in an effective and efficient way. On the contrary, blocking techniques exploit a simple similarity, based on common tokens, rather than the Jaccard similarity metric.[2] Such techniques seem to be more resilient to the inherent diversity and incompleteness of entity descriptions published in the Web of data. Alternatively, as we will see in Chapter 4, LSH techniques have been proposed as a pre-processing step for iterative approaches, in which the missed matches of an iteration can be identified in a subsequent iteration.

---

[2]Recent works extend LSH for non-metric spaces [Mu and Yan, 2010], while others also address distributed LSH computation [Bahmani et al., 2012, Silva et al., 2014].

CHAPTER 3

Blocking

As we have seen in Chapter 2.1, grouping entity descriptions in blocks before comparing them for matching is an important pre-processing step for pruning the quadratic number of comparisons required to resolve a collection of entity descriptions. The main objective of algorithms for entity blocking, formally defined in Section 3.1, is to achieve a reasonable compromise between the number of comparisons suggested and the number of missed entity matches. In Section 3.2, we briefly present traditional blocking algorithms proposed for relational records and explain why they cannot be used in the Web of data. Then, in Section 3.3, we detail a family of algorithms that relies on a simple inverted index of entity descriptions extracted from the tokens of their attribute values. Hence, two descriptions are placed into the same block if they share at least a common token. As we will see in Section 3.4, a more precise similarity (e.g., Jaccard) comparison of two entity descriptions can be achieved by post-processing the blocks of the inverted index and thus further reduce the number of entity pairs that need to be compared.

3.1 THE PROBLEM OF ENTITY BLOCKING

Blocking can be seen as an indexing technique, which places similar entity descriptions into blocks. After blocking, each description has to be compared only to others placed within the same block and thus disregard comparisons between descriptions that are unlikely to be matches. The two main desiderata of blocking techniques are to place (i) similar descriptions in the same block, aiming at *effectiveness*, and (ii) dissimilar descriptions in different blocks, aiming at *efficiency*. Clearly, it is not easy to accomplish simultaneously both (i) and (ii) since in general, efficiency dictates skipping many comparisons, possibly leading to many missing matches, which in turn implies low effectiveness. This is even more critical in the context of the Web of data, in which we do not know which pieces of the descriptions are the most appropriate to consider for computing the similarities. Hence, in practice, we are interested in maximizing the intersection of the Venn diagram depicted in Figure 2.3 (i.e., true matches), while keeping a reasonable size of the difference with the set of matching pairs (i.e., false matches).

Given a set of entity descriptions $\mathcal{E}$, we formally define a blocking collection as a set of blocks containing the descriptions in $\mathcal{E}$.

Definition 3.1 Blocking collection. Let $\mathcal{E}$ be a set of entity descriptions. A blocking collection is a set of blocks containing entity descriptions, $B = \{b_1, \ldots, b_m\}$, such that, $\bigcup_{b_i \in B} b_i = \mathcal{E}$.

In general, blocking techniques are characterized by their *redundancy attitude* as: (i) *partitioning*, that place each description into a single block, and (ii) *overlapping*, that could place a description in multiple blocks. Following a partitioning approach, a wrong decision on the block in which a description is placed would directly result in missed matches, if such exist. On the other hand, placing entity descriptions in multiple blocks, as in overlapping approaches, reduces the chances of missing true matches, but entails a greater number of comparisons. As a matter of fact, the occurrence of two descriptions in several blocks, provides evidence regarding their similarity [Papadakis et al., 2014a]. This way, overlapping approaches can be further divided into: (a) *overlap-positive*, that consider the number of common blocks between two descriptions proportional to the likelihood that they are matches, (b) *overlap-negative*, that consider the number of common blocks between two descriptions inversely proportional to the likelihood that they are matches, and (c) *overlap-neutral*, that consider the number of common blocks between two descriptions irrelevant to the likelihood that they are matches.

3.2   BLOCKING IN TRADITIONAL DATA WAREHOUSES

Broadly, traditional blocking techniques proposed for relational data can be distinguished between hash-based and sort-based. *Hash-based* techniques focus on mapping entity descriptions to blocks, taking into account specific criteria on the descriptions data. That is, descriptions are placed into blocks without performing any comparisons between them. *Sort-based* approaches arrange descriptions according to a certain sequence; blocking is performed then based on this arrangement, again without performing any comparisons. Traditionally, both approaches rely on the existence of *blocking keys*, i.e., constraints on a fixed set of attributes, based on which of the descriptions are placed into blocks.

A typical hash-based blocking [Fellegi and Sunter, 1969] has been originally proposed for tabular data. Given a blocking key, the block in which a description will end up is determined by a similarity function[1] on the value of the description for the blocking key. Such a value is called *blocking key value* (BKV). This way, standard blocking produces partitions of entity descriptions by putting descriptions with the same BKV into the same block. So, each distinct pair of descriptions is compared at most once, since each description is placed in exactly one block. By considering several blocking keys, we can potentially generate many BKVs for each description, and thus place it in more than one block.

Following the intuition of the overlap-positive approaches, [Gravano et al., 2001] creates multiple BKVs for a description, by converting each initial BKV into a list of q-grams, where a *q-gram* is a substring of q characters. Sub-lists of this list are generated, by recursively removing one q-gram at a time. Each sub-list is then converted (by concatenation) into a string and used as a BKV. Descriptions with a common BKV are placed in the same block. This way, typographical, or spelling errors are excused.

---

[1]Similarity functions in this setting usually overcome data glitches in attribute values such as typographical, or spelling errors.

Example 3.2   The string "Eiffel," in the q-gram based blocking approach, can be converted to the list of bi-grams ["ei,"'if,"'ff,"'fe,"'el"]. Some of the sub-lists for "Eiffel" are ["ei,"'if,"'ff," fe,"'el"], ["if,"'ff,"'fe,"'el"], ["ei," ff,"'fe,"'el"], and ["ei,"'ff,"'el"]. So, descriptions with the initial BKVs "Eiffel" and "Eifel," respectively, will be placed in some common blocks.

In a similar way, *suffixes* of BKVs, i.e., sub-strings produced by removing some of the first characters of the BKVs, can be used for blocking [Aizawa and Oyama, 2005]. Each suffix corresponds to a distinct block, and entity descriptions containing this suffix are inserted into this block by ignoring potential errors in the removed characters. To prevent a large number of descriptions being placed into the same block, two thresholds are set: (i) a threshold reflecting the minimum length of suffix strings that will be generated, and (ii) a threshold reflecting the maximum block size, i.e., number of entity descriptions contained in each block.

An alternative approach is to somehow sort entity descriptions and perform blocking based on the resulting order. The underlying assumption is that matching descriptions will become neighbors after the sorting, and thus neighbor descriptions constitute candidate matches. [Hernàndez and Stolfo, 1995] presents a *sorted neighborhood* method that works as follows. Initially, entity descriptions are ordered based on their BKV. Then, a window, resembling a block, of fixed length slides over the ordered descriptions, each time comparing only the contents of the window. An adaptive variation of this method is to dynamically decide on the size of the window [Yan et al., 2007]. In this case, adjacent BKVs in the sorted descriptions that are significantly different from each other, are used as boundary pairs, marking the positions where one window ends and the next one starts. Hence, this variation creates non-overlapping blocks.

Traditional blocking in MapReduce. Blocking, even as a pre-processing step for entity resolution, is a heavy computational task. This way, several approaches exploit the MapReduce programming model [Dean and Ghemawat, 2008] for parallelizing blocking algorithms (see Chapter 3.2 of [Dong and Srivastava, 2015] for an overview of algorithmic techniques). Abstractly, a collection of entity descriptions, given as input to a MapReduce job, is split into smaller chunks, which are then processed in parallel. A *map* function, emitting intermediate (*key*, *value*) pairs for each input split, and a *reduce* function that processes the list of *values* of an intermediate *key*, coming from all the mappers, should be defined. [Kolb et al., 2012b] provides a MapReduce implementation for hash-based blocking. In the map phase, for each description, a (*BKV*, *description*) pair is emitted. Descriptions with the same BKV are assigned to the same reduce task. Thus, each reduce task receives a block of descriptions and performs comparisons only between them. The MapReduce implementation of a sorted neighborhood blocking is given in [Kolb et al., 2012a]. This implementation assumes an ordering of descriptions and requires comparing descriptions within sliding windows that spread over different reduce tasks, while the map phase replicates descriptions close to partition boundaries and forwards them to both the respective reduce task and its successor.

Blocking techniques, thoroughly studied for relational data, assume both the availability and knowledge of the schema of the data. As a result, they cannot be used for the Web of data, where we do not even know on which attributes entity descriptions should be compared for similarity.

3.3   BLOCKING IN THE WEB OF DATA

To support a Web-scale resolution of heterogeneous and loosely structured entities across domains, blocking algorithms disregard strong assumptions about knowledge of the schema of data and rely on a minimal number of assumptions about how entity descriptions match within or across KBs.

Token blocking. Token blocking [Papadakis et al., 2011a] is a simple approach that relies on the minimal assumption that matching descriptions should at least share a common token. In this hash-based method, each distinct token t in the set of values of an entity description defines a new block b_t. Token blocking builds essentially an inverted index of entity descriptions: each token is a key and each block forms a list that is associated with this key. Two descriptions will be placed in the same block, if they share a token in their sets of values. Consequently, each description will be placed into multiple blocks according to its number of tokens.

Example 3.3   Given a single dirty entity collection consisting of the descriptions of Figure 3.1, Figure 3.2 shows the generated blocks after applying token blocking on the descriptions. The pair (e_1, e_6) is contained in four different blocks, which typically leads to comparing these descriptions four times. Moreover, a number of pairs, such as (e_1, e_2), (e_4, e_6), (e_3, e_7) and many others, leads to performing redundant comparisons, that is, comparisons that do not return matching descriptions.

e_1 = {(about, Eiffel Tower), (architect, Sauvestre), (year, 1889), (located, Paris)}
e_2 = {(about, Statue of Liberty), (architect, Bartholdi Eiffel), (year, 1886), (located, NY)}
e_3 = {(about, Auguste Bartholdi), (born, 1834), (work, Paris)}
e_4 = {(about, Joan Tower), (born, 1938)}
e_5 = {(work, Lady Liberty), (artist, Bartholdi), (location, NY)}
e_6 = {(work, Eiffel Tower), (year-constructed, 1889), (location, Paris)}
e_7 = {(work, Bartholdi Fountain), (year-constructed, 1876), (location, Washington)}

Figure 3.1: A set of entity descriptions.

Token blocking offers a brute-force method that allows comparing entity descriptions, even if they are considerably heterogeneous. In the remaining of this section, we will present two extensions of token blocking, namely attribute clustering blocking, in which candidate for matching descriptions should at least share a common token, only for similar attributes known globally, and prefix-infix(-suffix) blocking, in which candidate for matching descriptions should additionally share a common URI infix.

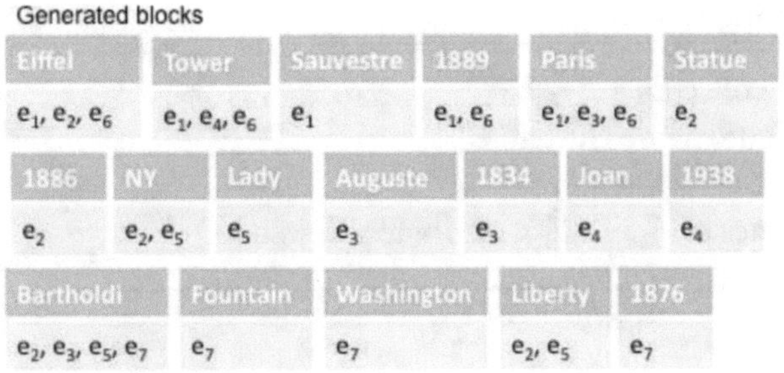

Generated blocks

Figure 3.2: Token blocking example.

Attribute clustering blocking. Attribute clustering blocking [Papadakis et al., 2013] exploits schematic information of the entity descriptions in order to minimize the number of false matches. To achieve this, prior to token blocking, it proposes an initial clustering of attributes based on the similarities of their values as observed over the entire collection of descriptions. Then, rather than looking for a common token regardless of the attribute it occurs, entity descriptions are compared only on the values of similar attributes. Hence, comparisons between descriptions that do not share a common token in a similar attribute are discarded.

In a clean-clean entity resolution scenario, Algorithm 1 shows how to group similar attributes. Every attribute of one collection (line 3) is connected to its most similar attribute of the other collection (line 6) and, based on the transitive closure of connected attributes (lines 12–13), attribute clusters are formed. Singleton clusters are merged to form a so-called glue cluster (lines 14–20). Then, for each attribute cluster, token blocking is performed. That is, each distinct token t in the values of an attribute belonging to a cluster c, defines a block $b_{c,t}$. This way, two entity descriptions will be placed in the same block, if they share a common token in their sets of values, for attributes of the same cluster. Like token blocking, attribute clustering blocking generates overlapping blocks. Compared to the blocks created by token blocking, it is expected to produce a larger number of smaller blocks.

Example 3.4   Consider two clean entity collections, $D_1 = \{e_1, e_2, e_3, e_4\}$ and $D_2 = \{e_5, e_6, e_7\}$, that together compose the descriptions of Figure 3.1. To find, for instance, the attribute of D_2 that is the most similar to the attribute *about* of D_1, we compute the similarities between the values of *about* (i.e., the values: Eiffel, Tower, Statue, Liberty, Auguste, Bartholdi, Joan) and the values of the attributes of D_2. Using Jaccard similarity, we deduce that the attribute *work* (with values: Lady, Liberty, Eiffel, Tower, Bartholdi, Fountain) of D_2 is the most similar attribute to *about* of D_1. In a similar fashion, we identify the pairs of the most similar attributes between D_1 and D_2 (Figure 3.3(a) depicts such pairs), based on which we produce the clusters of attribute names (Figure 3.3(b)). A subset of the blocks constructed for each cluster of attribute names is

---

Algorithm 1 Attribute-clustering

Input : Clean entity collections E_1, E_2
Output : Set of attribute name clusters C

$N(E_i)$: attribute names of E_i, $V(E_i)$: attribute values of E_i

---

1: $links \leftarrow \emptyset$;
2: $c_{glue} \leftarrow \emptyset$;
3: **for all** $n_{i,1} \in N(E_1)$ **do**
4:     $n_{j,2} \leftarrow getMostSim(n_{i,1}, N(E_2), V(E_2))$;
5:     **if** $sim(n_{i,1}.getVal(), n_{j,2}.getVal()) > 0$ **then**
6:         $links.add(n_{i,1}, n_{j,2})$;
7:     **end if**
8: **end for**
9: **for all** $n_{i,2} \in N(E_2)$ **do**
10:     ...//same as with $N(E_1)$
11: **end for**
12: $links' \leftarrow transitiveClosure(links)$;
13: $C \leftarrow getConnectedComponents(links')$;
14: **for all** $c_i \in C$ **do**
15:     **if** $|c_i| = 1$ **then**
16:         $C.remove(c_i)$;
17:         $c_{glue}.add(c_i)$;
18:     **end if**
19: **end for**
20: $C.add(c_{glue})$;
21: **return** C;

---

shown in Figure 3.3(c). Attribute clustering offers higher efficiency than token blocking, i.e., it produces less comparisons, but again, many non-matches may be placed in the same block (e.g., e_4 and e_6 are both placed in block $C1.Tower$ (Figure 3.3(c))), and same pairs of descriptions could be contained in many blocks (e.g., e_1 and e_6 are placed together in four different blocks).

Prefix-infix(-suffix) blocking. From a different viewpoint, prefix-infix(-suffix) blocking [Papadakis et al., 2012] exploits information regarding the names of the entity descriptions published on the Web. Its basic assumption is that matching descriptions have a common token in their literal values, or a common URI.

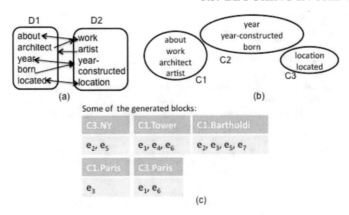

Figure 3.3: Attribute clustering blocking example: (a) most similar attribute-pairs, (b) attribute clusters, (c) generated blocks.

Specifically, given that many URIs follow a naming policy, they can be used as evidence regarding possible matching descriptions. In [Papadakis et al., 2010], it is measured that approximately 66% of the 182 million URIs of the BTC09 data set[2] follow the prefix-infix(-suffix) pattern. The prefix describes the source, i.e., domain, of the URI, the infix is a local identifier, and the optional suffix contains either details about the format, e.g., .rdf and .nt, or a named anchor.

Example 3.5 For example, in the URI "http://liris.cnrs.fr/olivier.aubert/foaf.rdf#me," the prefix is "http://liris.cnrs.fr," the infix is "/olivier.aubert," and the suffix is "/foaf.rdf#me."

Several variations of this blocking technique were proposed. Given a set of entity descriptions, the best-performing blocking creates one block for each distinct token in the literal values of the descriptions and one block for each distinct infix in the URIs of the descriptions (i.e., the name of the described entity or any other neighborhood entity). In case an infix consists of a single token that appears in a literal, we do not construct a new block for it. This approach is of course constrained by the extent to which common naming policies are followed by the KB publishers. Compared to token blocking, in a favorable scenario, it is expected to create additional blocks for the names of the descriptions, which enables us to consider matching descriptions, even when they have no common tokens in their literal values.

Example 3.6 Figure 3.4(c) shows the blocks produced after applying prefix-infix(-suffix) blocking to the entity descriptions of Figure 3.4(a) (actually, the descriptions of Figure 3.4(a) are the descriptions of Figure 3.1, slightly modified to exploit and understand the characteriztics of the

[2]http://km.aifb.kit.edu/projects/btc-2009/

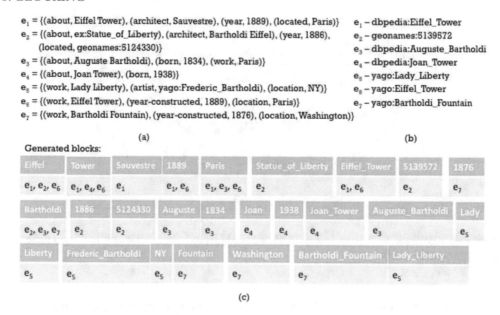

Figure 3.4: Prefix-infix(-suffix) blocking example: (a) the input entity collection, (b) URI identifiers of the descriptions, (c) generated blocks.

method, while Figure 3.4(b) presents the URIs that correspond to the identifiers of the entity descriptions).

Join-based blocking. An alternative approach for blocking is to consider *string-similarity join* algorithms. In a nutshell, such algorithms construct blocks by identifying all pairs of descriptions whose string values similarities are above a certain threshold and potentially some pairs whose string values similarities are below that threshold. To achieve that, without computing the similarity of all pairs of descriptions, string-similarity join algorithms (e.g., [Bayardo et al., 2007, Chaudhuri et al., 2006, Xiao et al., 2011]) build an inverted index from the tokens of the descriptions values. However, unlike token blocking, this inverted index is created only by the first few non-frequent tokens of each description (i.e., the most discriminating), based on the *prefix filtering* principle [Chaudhuri et al., 2006]. This principle states that, if the p-prefix of token set x is the first p tokens of x, two token sets x and y with an intersection size $|x \cap y| \geq t$ should at least share a common token in the $(|x| - t + 1)$-prefix of x and the $(|y| - t + 1)$-prefix of y. Moreover, based on the observation that

$$Jaccard(x, y) \geq t \Rightarrow |x \cap y| \geq \frac{t}{1+t} \cdot (|x| + |y|), \tag{3.1}$$

we only need to index a prefix of length $|x| - \lceil t \cdot |x| \rceil + 1$, for every token set x, to ensure the prefix filtering-based method does not miss any similar pair, i.e., any pair of token sets with Jaccard similarity t or greater. To avoid generating large blocks originating from stop words, the tokens within each description are sorted in ascending order of document frequency, namely, stop words are placed last. [Bayardo et al., 2007] additionally applies a *size filtering* [Arasu et al., 2006] on the sets of tokens to disregard some of the candidate pairs, based on the fact that $Jaccard(x, y) \geq t \Rightarrow t \cdot |x| \leq |y|$.

Furthermore, the ppjoin+ algorithm [Xiao et al., 2008, 2011] exploits prefix- and size-filtering, also introducing a *positional filtering*, i.e., the position in the ordered set of tokens, in which a token appears, to further reduce the number of candidate pairs. Specifically, it estimates the maximum possible intersection size of two token sets x, y, by considering that if the first common token of x and y is the first token of x and the second token of y, then the maximum intersection that these sets can have is $1 + min(|x| - 1, |y| - 2)$. For a complete survey and experimental evaluation of string-similarity join algorithms, we refer the reader to [Augsten and Böhlen, 2013, Jiang et al., 2014].

Example 3.7   Assume that we want to find all the pairs (e_i, e_j) of descriptions of Figure 3.1 with $Jaccard(e_i, e_j) \geq 0.7$, considering only the sets of tokens in the values of these descriptions. These descriptions can be transformed to (tokens in ascending order of frequency):

e_1 = {Sauvestre, 1889, Eiffel, Tower, Paris},
e_2 = {Bartholdi, 1886, Eiffel, Liberty, Statue, NY, Of},
e_3 = {1834, Bartholdi, Auguste, Paris},
e_4 = {1938, Tower, Joan},
e_5 = {Bartholdi, Liberty, Lady, NY},
e_6 = {1889, Eiffel, Tower, Paris},
e_7 = {Bartholdi, 1876, Fountain, Washington}.

If we only index the $|x| - \lceil t \cdot |x| \rceil + 1$ first tokens (underlined) of each token set x, then the resulting blocks will be those of Figure 3.5. The candidate pairs will be (e_1, e_6), (e_2, e_3), (e_2, e_5), (e_2, e_7), (e_3, e_5), (e_3, e_7), (e_5, e_7), (e_2, e_6). Out of those, only (e_1, e_6) has a Jaccard similarity above the pre-defined threshold, while the matching pair (e_2, e_5) has a much lower Jaccard similarity (0.375). Additional filtering can be applied to disregard some of the candidate pairs, based on the size- and positional-filtering principles. For example, by applying size-filtering, we induce that it would not be possible for the pair (e_2, e_3) to have a Jaccard similarity of 0.7, since $|e_3| < 0.7 \cdot |e_2|$. The same principle can also be applied to disregard the comparisons between the candidate pairs (e_2, e_5), (e_2, e_7), (e_2, e_6). By applying positional-filtering, we can also disregard the pair (e_3, e_5), since their first common token is the second token of e_3 and the first token of e_5, meaning that their maximum intersection size is $1 + min(|e_3| - 2, |e_5| - 1) = 3$, while based on Equation (3.1), their intersection should be at least 4.7.

Tuning the similarity threshold required by join-based algorithms is non-trivial and it also affects the performance of such algorithms [Jiang et al., 2014]. Smaller thresholds entail less

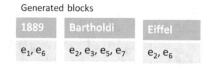

Figure 3.5: The blocks generated by a set-similarity join method for the descriptions of Figure 3.1.

pruning, and thus, more time. Furthermore, [Metwally and Faloutsos, 2012] proves experimentally that algorithms based on prefix filtering are only effective when the similarity threshold is extremely high. However, this is not the case in the Web of data, in which highly heterogeneous descriptions, i.e., yielding very low similarity in their literal values, can refer to the same entity.

Blocking based on frequent itemsets. In the previous methods, the constructed blocks represent one token. A method to reduce the number of compared descriptions consists of building blocks for sets of tokens that appear together in many entity descriptions (i.e., frequent itemsets [Rajaraman and Ullman, 2011]). Several variations of this problem have been proposed. [Miliaraki et al., 2013] studies the problem of scalable finding frequent sets of tokens that appear in specific sequences. [Kenig and Gal, 2013] introduces a technique for building blocks based on the maximal frequent itemsets identified by [Grahne and Zhu, 2005]. Abstractly, each maximal frequent itemset defines a block, and descriptions containing the tokens of a frequent itemset are placed in the block that the itemset defines. Using frequent itemsets to construct blocks may significantly reduce the number of candidate-for-matching pairs. However, ignoring non-frequent sets of tokens may also significantly increase the chances of missing matches, especially between descriptions with few common tokens.

Multidimensional blocking. Finally, [Isele et al., 2011] proposed the concept of *multidimensional overlapping* blocks. Focusing on resolving entity descriptions based on a set of similarity functions, this method firstly constructs a collection of blocks for each similarity function. This way, for a specific function, similar descriptions will be placed into the same block in the collection corresponding to this function. Then, all blocking collections are aggregated into a single multidimensional collection, taking into account the similarities of descriptions that share blocks. Multidimensional blocking has been implemented in the context of the Silk Link Discovery Framework [Volz et al., 2009] that targets at identifying links between descriptions.

Token blocking and variations in MapReduce. To cope with the large volume of entity descriptions published in the Web of data, parallel implementations of blocking algorithms are required. Next, we provide such implementations in MapReduce. Regarding token blocking, for each description e, a (t, e) pair is emitted by the mapper for each token t in the set of values of e. In the reduce phase, descriptions having a common token will be processed by the same reduce task, i.e., they are placed in the same block. A similar procedure is also used, for example, in [McCreadie et al., 2012] for constructing inverted indices.

Given two clean entity collections, attribute clustering blocking can be briefly sketched by the following steps, each representing a MapReduce job:

- *Attribute Creation.* In the first job, we create a list of values for each unique attribute, in each collection, appearing in the descriptions of the collection.

- *Attribute Similarities.* In the second job, we compute the similarities between all attribute pairs, containing attributes from different collections. To do that, we compare attributes of each data chunk to each other, as well as to attributes of all the other chunks, similarly to the non-approximate algorithm of [Zhang et al., 2012a].

- *Best Match.* In the third job, we keep for each attribute of each collection, only the attribute of the other collection with the highest similarity score.

- *Clustering and Blocking.* In the final step, we associate each attribute with a cluster id. Then, we perform token blocking, except that in each *key* emitted from a mapper, there is also a cluster prefix, enabling distinctions between blocks for the same token.

Given the infixes of the entity descriptions as input (for an infix extraction algorithm, see [Papadakis et al., 2010]) and the descriptions themselves, the MapReduce implementation of prefix-infix(-suffix) blocking uses two different mappers, operating in parallel. The first mapper is the one used in token blocking, applied only in the literal values of the input descriptions. The second mapper forwards to the reducer blocks with (*key*, *value*) pairs, where *key* is an infix and *value* is a description having this infix. In the reduce phase, all the descriptions having a common token or infix in their literals or URIs will be placed in the same block.

Finally, ppjoin+ is adapted in MapReduce in [Vernica et al., 2010]. This adaptation relies on three phases. First, it produces a list of join tokens ordered by frequency count. Next, the descriptions along with their join tokens are extracted from the dataset, distributed to the reduce tasks, which in turn produce pairs of similar descriptions. The third phase eliminates the repeated description pairs.

3.4   BLOCK POST-PROCESSING METHODS

Most of the blocking methods are overlap-positive in the sense that their blocks provide positive evidence for the matching likelihood of two descriptions: the more blocks the descriptions share, the more likely they are to match. The main characteriztic of overlap-positive blocking methods is that they trade very high recall for very low precision. Or, in other words, they yield a large number of suggested comparisons in their effort to achieve high recall. In this respect, block post-processing methods attempt to reduce (i) *redundant comparisons*, i.e., comparisons between descriptions that have already been considered in a previously examined block, (ii) *unnecessary comparisons*, i.e., comparisons between descriptions of different entity collections that have already been found matching to another description and thus cannot produce new matches (in a clean-clean entity resolution task), and (iii) *comparisons between unlikely-to-match* descriptions. [Papadakis et al., 2013] proposes different block post-processing methods for each of the above cases, which could also rely on an appropriate ordering of candidate for matching descriptions. Clearly, the elimination of redundant and unnecessary comparisons decreases the number of performed comparisons, while it maintains the recall of the original blocking method. However, the order by which candidate pairs are examined affects the estimation for unlikely-to-match descriptions and thus the overall recall of the blocking process.

For example, [Papadakis et al., 2011b] proposes a method for discarding all redundant comparisons from any set of blocks. In essence, when two descriptions are compared in a block, this comparison is not performed again in any other block this pair appears. [Papadakis et al., 2011a] proposes a method for reducing unnecessary comparisons in a clean-clean entity resolution task. To do this, before comparing a pair of descriptions, we examine whether any of them has been previously matched with a third description. In [Papadakis et al., 2011a, Whang et al., 2013b], blocks are ordered based on a utility function (e.g., their size). The underlying assumption is that blocks placed at the highest ranking positions are more likely to contain more matches. Thus, low-ordered blocks are removed and the rest are examined in this specified order, until a good trade-off between recall and reduction on the number of generated comparisons has been reached.

Example 3.8   Given the blocks of Figure 3.2 (excluding blocks containing a single description) a block post-processing method does the following. First, it sorts the blocks in ascending order of size, as depicted in Figure 3.6 (top). Then, it removes oversized blocks (e.g., containing more than three descriptions, like the block corresponding to the token *Bartholdi*). For each entity description, it builds an index of blocks, in which this description appears, sorted according to the relative order of the blocks, as shown in Figure 3.6 (bottom). Finally, it discards comparisons that have been already suggested in a previous block, e.g., the comparisons between (e_2, e_5) is discarded for the block *Liberty*, since it has been already suggested by the block *NY*, as found by the indices of e_2 and e_5 (they have two blocks in common, *Liberty* and *NY*, and *Liberty* appears after *NY*).

Meta-blocking. More recently, [Papadakis et al., 2014a] proposes to reconstruct the blocks of a given blocking collection (unlike the previous block post-processing methods) in order to more drastically discard redundant comparisons, as well as comparisons between descriptions that are unlikely to match. Meta-blocking essentially transforms a given blocking collection B into the *blocking graph* G_B. Its nodes V_B correspond to the descriptions in B, while its undirected edges E_B connect the co-occurring descriptions. No parallel edges are allowed, thus eliminating all redundant comparisons. Every edge $e_{i,j}$ is associated with a weight $w_{i,j} \in [0, 1]$ representing the likelihood that the adjacent entities are matching candidates. The low-weighted edges are pruned, so as to discard comparisons between unlikely-to-match descriptions.

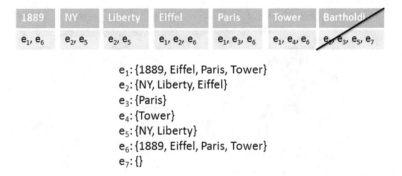

e₁: {1889, Eiffel, Paris, Tower}
e₂: {NY, Liberty, Eiffel}
e₃: {Paris}
e₄: {Tower}
e₅: {NY, Liberty}
e₆: {1889, Eiffel, Paris, Tower}
e₇: {}

Figure 3.6: The blocks of Figure 3.2 in ascending order of size (top) and the corresponding entity index (bottom).

Different *weighting schemes* for edges have been proposed. More precisely, the weight of an edge connecting two descriptions can be the number of common blocks shared by these descriptions, or the Jaccard similarity of the sets of blocks that contain these descriptions, or the inverse of the sum of candidate pairs contained in the common blocks of these descriptions. These weighting schemes estimate the likelihood that two descriptions are potentially matching based on the number of their common tokens eventually normalized by their frequency (reflected in the size and number of common blocks they appear). Finally, different *pruning strategies* exploit the information provided by a blocking graph. Two pruning criteria can be used, namely, the minimum weight of the retained edges and the maximum number of retained edges. With respect to its application scope, the pruning criterion can be either global, applying to the entire blocking graph, or local, covering only a node neighborhood.

Example 3.9 The blocks in Figure 3.2, produced after applying token blocking to the descriptions of Figure 3.1, can be mapped to the blocking graph shown in Figure 3.7(a). For simplicity here, we consider that each edge weight is equal to the number of blocks shared by its adjacent descriptions. Then, different algorithms can be used to remove edges with low weights and discard part of the unnecessary comparisons. For instance, one policy is to discard all edges having a weight lower than the average edge weight across the entire graph. For the blocking graph of Figure 3.7(a), the average edge weight is 1.385. The resulting pruned blocking graph is shown in Figure 3.7(b). Finally, meta-blocking outputs a blocking collection generated from this pruned graph by placing the adjacent entities of every edge into a separate block. In overall, the new collection contains just two comparisons, unlike the initial collection containing eighteen comparisons, and does not miss any of the two matches.

Moving forward, [Papadakis et al., 2014b] formalizes meta-blocking as a binary classification task, targeting at identifying edges that correspond to matches and non-matches between their adjacent entity descriptions. Instead of assigning unilateral weights to the edges (as in Fig-

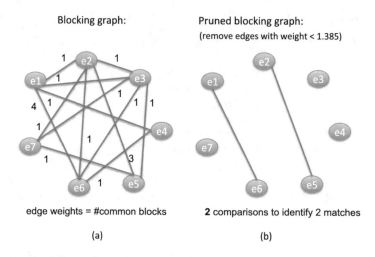

Figure 3.7: Meta-blocking example: (a) depicts a blocking graph, which is pruned (b), to discard unnecessary comparisons.

ure 3.7), this method composes information about the co-occurring entities into comprehensive feature vectors. Such features can be, for example, the number of common blocks between the two descriptions, or the number of descriptions contained in their common blocks. The resulting feature vectors are fed into a supervised classification algorithm that learns composite rules, instead of the simple rules of the form "if weight < threshold then discard edge", to effectively distinguish matching and non-matching edges based on small, manually created training sets. Concerning the set of features annotating the edges of the blocking graph, using more features may help make the pruning of the non-matching edges more accurate. However, the computational cost for meta-blocking gets higher. Therefore, a small set of generic features that combine a low extraction cost with high discriminatory power are suggested.

Example 3.10   Consider that each edge of the blocking graph is associated with a feature vector $[a_1, a_2]$ (not a single weight as in Figure 3.7 (a)), where a_1 is the number of common blocks shared by the adjacent descriptions, and a_2 is the total number of comparisons contained in these blocks. A composite rule (unlike the simple one used in Figure 3.7 (b)) could be "if $a_1 \leq 2$ and $a_2 > 5$ then discard edge", capturing the intuition that the more blocks two descriptions share and the smaller these blocks are, the more likely the descriptions match.

3.5   DISCUSSION

Given k KBs with d entity descriptions each, a brute-force entity resolution approach, without using blocking, requires $O(d^k)$ comparisons. This is quickly prohibitive for even moderate k or

Table 3.1: Criteria for placing descriptions in the same block

Method	Criterion
Token Blocking [Papadakis et al., 2011a]	The descriptions have a common token in their values.
Attribute Clustering Blocking [Papadakis et al., 2013]	The descriptions have a common token in the values of attributes that have similar values in overall.
Prefix-Infix(-Suffix) Blocking [Papadakis et al., 2012]	The descriptions have a common token in their literal values, or a common URI infix.
ppjoin+ [Vernica et al., 2010, Xiao et al., 2008]	The descriptions have a common token in their p first tokens; tokens are sorted in ascending order of frequency.
Frequent itemsets [Kenig and Gal, 2013]	The descriptions have the tokens of a frequent itemset in their values.

Table 3.2: Blocking approaches with respect to the redundancy attitude and algorithmic attitude

Blocking approach	Partitioning	Overlap-positive	Overlap-negative	Overlap-neutral	Hash-based	Sort-based
Standard blocking [Fellegi and Sunter, 1969, Kolb et al., 2012b]	✓				✓	
Q-grams [Gravano et al., 2001]		✓			✓	
Suffixes [Aizawa and Oyama, 2005]		✓			✓	
Sorted neighborhood [Hernàndez and Stolfo, 1995, Kolb et al., 2012a]				✓		✓
Adaptive sorted neighborhood [Yan et al., 2007]	✓					✓
Token blocking [Papadakis et al., 2011a]		✓			✓	
Attribute clustering blocking [Papadakis et al., 2013]		✓			✓	
Prefix-infix(-suffix) blocking [Papadakis et al., 2012]		✓			✓	
ppjoin+ [Vernica et al., 2010, Xiao et al., 2008]		✓			✓	
Frequent itemsets [Kenig and Gal, 2013]		✓			✓	

The header of Table 3.2 groups columns as: **Redundancy attitude** (Partitioning; Overlapping: Overlap-positive, Overlap-negative, Overlap-neutral) and **Algorithmic attitude** (Hash-based, Sort-based).

d. Instead, when with blocking, entity resolution is only applied within blocks, reducing the comparisons to $O(d_{b_{max}}^k)$, with $d_{b_{max}}$ being the size of the largest block.

Clearly, blocking techniques proposed for traditional data warehouses cannot be applied to resolve entities in the Web of data. First, they consider that two descriptions commit to the same relational schema (in most of the cases, they are seen as rows of the same table) and second, because their similarity is computed using known blocking keys. Given the high heterogeneity, loose structuring and varying quality of entity descriptions in the Web of data, we need blocking techniques that rely on a minimal number of assumptions about how descriptions match within or across KBs. Table 3.1 summarizes, simplified, the criteria employed by blocking techniques proposed for Web data in order to place two descriptions into the same block. The categorization of the blocking techniques presented in this lecture with respect to the characteriztics of the produced blocks (i.e., partitioning vs. overlapping blocks) and the algorithmic approach (hash-based vs. sort-based) used are presented in Table 3.2. Partitioning approaches are sensitive to errors, since misplaced entity descriptions potentially result in missed matches. Therefore, due to the varying data quality, they are not suited for entity resolution in the Web of data. Data heterogeneity makes sort-based approaches not easily applicable as well, since the missing knowledge of the schema of the data incommodes the sorting process.

Overall, existing blocking and block post-processing methods for the Web of data rely on the tokens in the values of the descriptions. The underlying assumption that such methods make is that matching descriptions should be identified based only on their values. However, as we will experimentally see in Chapter 5, information extracted from the graph-structure of the descriptions can be also exploited, in order to identify more matches. Toward this direction, blocking and matching phases can be employed in an iterative fashion (see Chapter 4). Ideally, to identify matches that non-iterative blocking algorithms miss, such iterative algorithms should take into account, on each iteration step, information stemming not only from the values, but also from the neighborhoods of the descriptions.

CHAPTER 4

Iterative Entity Resolution

As we have seen in Chapter 2.1, to minimize the number of missed matches, an iterative entity resolution (ER) process can progressively exploit any intermediate results of blocking and matching, discovering new candidate description pairs for resolution, even if this process entails additional processing cost. The main objective of the algorithms for iterative entity resolution, which is abstractly described in Section 4.1, is to identify matches based on knowledge gained from previously identified matches. We distinguish between merging-based (Section 4.2) and relationship-based (Section 4.3) iterative ER approaches. In the former, new matches can be identified by exploiting the merging of the previously located matches, while in the latter, iterations rely on the similarity evidence provided by descriptions being structurally related in the original entity graph. As we will see in Section 4.4, iterative ER can be also interleaved with the process of blocking, where matches are sought only within a block and if identified, they are propagated to other blocks. Finally, in Section 4.5, we overview works on incremental ER, in which the obtained ER results at each phase are enriched when new descriptions are made available (eventually in streams), and in Section 4.6, we present recent works on progressive ER, which attempt to discover as many matches as possible given limited computing budget, by estimating the matching likelihood of yet unresolved descriptions, based on the matches found so far.

4.1 THE PROBLEM OF ITERATIVE ENTITY RESOLUTION

The inherent distribution of entity descriptions in different KBs along with their significant semantic and structural diversity yield incomplete evidence regarding the similarity of descriptions published in the Web of data. Iterative ER approaches aim to tackle this problem by exploiting any partial result of the ER process in order to generate new candidate pairs of descriptions not considered in a previous step or even to revise a previous matching decision. From a different point of view, iterative approaches come to accomplish the need for fair partial results within a limited time period. As an example, consider real-time applications that cannot tolerate any entity resolution process that takes longer than a certain amount of time. For such cases, the entity resolution process can run iteratively, so as to discover additional matches in a pay-as-you-go fashion; the more the available time, the more the iterations for entity resolution, and so, the more the identified matches.

Abstractly, new matches can be found by exploiting merged descriptions of previously identified matches or relationships between entity descriptions. We call the iterative ER approaches that build their iterations on the merging of descriptions *merging-based*, and those that use entity

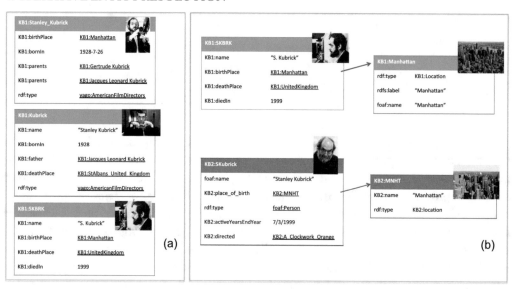

Figure 4.1: A merging-based iterative ER example (a) and a relationship-based iterative ER example (b).

relationships for their iteration step *relationship-based*. Intuitively, the merging-based approaches deal with descriptions of the same type, e.g., all descriptions refer to buildings, while relationship-based approaches presume upon the relationships between different types of entities.

Example 4.1   Consider the descriptions in Figure 4.1 (a), stemming, for example, from the knowledge base *KB1*, all referring to the person Stanley Kubrick. Initially, it is difficult to match *KB1:SKBRK* with any of the other descriptions, since many people named Kubrick may have been born in Manhattan, or died in the UK, respectively. However, it is quite safe to match the first two descriptions (*KB1:Stanley_Kubrick* and *KB1:Kubrick*). By merging the first two descriptions, e.g., using the union of their attribute-value pairs, it now becomes easier to identify that the last description (*KB1:SKBRK*) is also referring to the same person, based on the name, and places of birth and death.

   Consider now the descriptions in Figure 4.1 (b), stemming, for instance, from the knowledge bases *KB1* and *KB2*. The descriptions on the left (*KB1:SKBRK* and *KB2:SKubrick*) represent Stanley Kubrick, while the descriptions on the right (*KB1:Manhattan* and *KB2:MNHT*) represent Manhattan, where Kubrick was born. Initially, it is difficult to identify the match between the descriptions on the left, based only on the common year of death and last name. However, it is quite straightforward to identify the match between the descriptions of Manhattan, on the right. Having identified this match, a relationship-based iterative ER algorithm would re-consider matching *KB1:SKBRK* to *KB2:SKubrick*, since these descriptions are additionally related, with the same kind of relationship (birth place), to the descriptions of Manhattan that were previously matched. Therefore, a relationship-based iterative ER algorithm would identify this new match in a second iteration.

Interestingly, there is an important difference between the two families of iterative approaches. Namely, which is the fact that triggers a new iteration in each family of iterative approaches? Next, we highlight this by exploring the general framework of iterative entity resolution approaches that are typically composed of an initialization and an iterative phase [Herschel et al., 2012]. The goal of the initialization phase is to create a queue capturing the pairs of entity descriptions that will be compared, or even the order for comparing these pairs. For example, such a queue can be constructed automatically by exhaustively computing the initial similarity of all pairs of descriptions, or can be handled manually by domain experts, who specify, for instance, which entities will be compared. In the iterative phase, we get a pair of descriptions from the queue, compute the similarity of this pair to decide if it is a match, and with respect to this decision, we potentially update the queue. Actually, the updates in this phase trigger the next iteration of entity resolution. In merging-based approaches, when two matching descriptions are merged, the pairs in the queue in which these descriptions participate are updated, replacing the initial descriptions with the results of their merging. New pairs may be added to the queue as well, suggesting the comparison of the new merged description to other descriptions. In relationship-based approaches, when related descriptions are identified as matches, new pairs can be added to the queue, e.g., a pair of building descriptions is added to the queue, when their architects are matching, or even existing pairs can be re-ordered. Ideally, the iterative phase terminates when the queue becomes empty.

4.2   MERGING-BASED ITERATIVE ENTITY RESOLUTION

In *merging-based iterative entity resolution*, the matching decision between two descriptions triggers a merge operation, which transforms the initial entity collection by adding the new, merged description and potentially removing the two initial descriptions. This change also triggers more updates in the matching decisions, since the new, merged description needs to be compared to the other descriptions of the collection. Intuitively, the final result of merging-based iterative entity resolution is a new set of descriptions which are the results of merging all the matches found in the initial entity collection. In other words, each real-world entity described in the input entity collection is represented by a single description in the resolution results and each description in the resolution results represents a distinct real-world entity from the input entity collection. More formally:

Definition 4.2   Merging-based entity resolution.   Let $\mathcal{E} = \{e_1, \ldots, e_m\}$ be a set of entity descriptions, $M : D \times D \rightarrow \{true, false\}$ be a boolean match function and $\mu : D \times D \rightarrow D$ be a partial merge function, applicable only to pairs of matching descriptions, where D is the domain of entity descriptions. A merging-based entity resolution of $\mathcal{E}$ is the smallest set of descriptions $\mathcal{E}'$, such that:

(i) $\forall e_i, e_j \in \mathcal{E} : M(e_i, e_j) = true, \exists e_k \in \mathcal{E}' : \mu(e_i, e_j) \preceq e_k$, and

(ii) $\forall e_k \in \mathcal{E}', \exists e_l \in \mathcal{E} : e_l \preceq e_k$,

where $e_i \preceq e_j$ means that e_j holds the same or more information than e_i, regarding the same real-world entity (note that $e_i \preceq e_i$). This definition for merging-based ER is compliant to Definition 2.2 of ER, presented in Chapter 2, if additionally, merging is applied to each partition.

Considering the functions of matching M and merging μ as black boxes, [Benjelloun et al., 2009] presents merging-based iterative ER strategies that minimize the number of invocations to these potentially expensive black boxes. Merged entity descriptions are considered as new entity descriptions, hence being possible matches to other descriptions in the collection. In the same line of work, [Benjelloun et al., 2007] introduces a family of algorithms that distribute the workload of merging-based ER across multiple processors. Since both works consider matching and merging as black boxes, [Benjelloun et al., 2009] introduces a set of desirable properties that, when satisfied by those functions, lead to higher efficiency. These properties, called $ICAR$ properties for short, are:

- *Idempotence:* $\forall e_i, M(e_i, e_i) = true$ and $\mu(e_i, e_i) = e_i$.

- *Commutativity:* $\forall e_i, e_j, M(e_i, e_j) = true \Leftrightarrow M(e_j, e_i) = true$ and $\mu(e_i, e_j) = \mu(e_j, e_i)$.

- *Associativity:* $\forall e_i, e_j, e_k,$ if $\mu(e_i, \mu(e_j, e_k))$ and $\mu(\mu(e_i, e_j), e_k)$ exist, then $\mu(e_i, \mu(e_j, e_k)) = \mu(\mu(e_i, e_j), e_k)$.

- *Representativity:* If $e_k = \mu(e_i, e_j)$, then for any e_l such that $M(e_i, e_l) = true$, we also have $M(e_k, e_l) = true$.

Regarding the match function, idempotence and commutativity have been already discussed in Chapter 2, as reflexivity and symmetry, respectively, while representativity extends transitivity, by also including the merge function. As a note, consider that if associativity does not hold, it becomes harder to interpret a merged description, since this description depends on the order in which the source descriptions were merged.

One of the algorithms exploiting the $ICAR$ properties is the R-Swoosh algorithm [Benjelloun et al., 2009], which operates as follows. A set $\mathcal{E}$ of entity descriptions is initialized to contain all the input descriptions. Then, at each iteration, a description e is removed from $\mathcal{E}$ and compared to each description e' of the, initially empty, set $\mathcal{E}'$. If e and e' are found to match, then they are removed from $\mathcal{E}$ and $\mathcal{E}'$, respectively, and the result of their merging is placed into

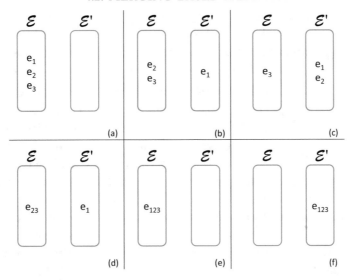

Figure 4.2: The execution of R-Swoosh for Example 4.3.

$\mathcal{E}$ (exploiting representativity). If there is no description e' matching with e, then e is placed in $\mathcal{E}'$. This process continues until $\mathcal{E}$ becomes empty, i.e., there are no more matches to be found.

Example 4.3   Consider the entity descriptions:

e_1 = {(name, Liberty), (architect, Bartholdi), (year-constructed, 1886)},

e_2 = {(about, Statue of Liberty), (architect, Eiffel), (location, NewYork)},

e_3 = {(name, Statue of Liberty), (architect, Bartholdi Eiffel), (year, 1886), (locatedIn, NewYork)}.

Abstractly, it is rather difficult to say if descriptions e_1 and e_2 are matching. However, it is quite safe to assume that e_2 and e_3 are matching. By merging these two descriptions into a new one, e.g.,

e_{23} = {(name, Statue of Liberty), (architect, Bartholdi Eiffel), (year, 1886), (location, NewYork)},

it becomes easier to deduce that e_1 is also matching with e_{23}.

Figure 4.2 illustrates the execution of the R-Swoosh algorithm for the three given entity descriptions. The set $\mathcal{E}$ initially contains e_1, e_2, and e_3, while the set $\mathcal{E}'$ is initialized as the empty set (Figure 4.2 (a)). In the first iteration, e_1 is removed from $\mathcal{E}$ and placed to $\mathcal{E}'$, since $\mathcal{E}'$ was previously empty (Figure 4.2 (b)). At the next iteration, e_2 is removed from $\mathcal{E}$ and compared to the descriptions in $\mathcal{E}'$, i.e., it is compared to e_1. However, their similarity is not high enough, so e_2 is also placed in $\mathcal{E}'$ (Figure 4.2 (c)). Next, e_3 is removed from $\mathcal{E}$ and compared to the contents of $\mathcal{E}'$, i.e., e_2 and e_1. When it is compared to e_2, they are found to be matching, so a new description e_{23}, which is the result of merging e_2 with e_3, is placed in $\mathcal{E}$ and e_2 is removed from $\mathcal{E}'$ (Figure 4.2 (d)). Next, e_{23} is removed from $\mathcal{E}$ and compared to the only description in $\mathcal{E}'$, i.e., e_1.

They are found to be matching, so they are merged as e_{123} and moved to $\mathcal{E}$, while e_1 is removed from $\mathcal{E}'$ (Figure 4.2 (e)). Finally, e_{123}, the only element of $\mathcal{E}$, is placed into $\mathcal{E}'$ (Figure 4.2 (f)), which is given as the result of resolving the given descriptions. Eventually, all the descriptions e_1, e_2, and e_3 are found to be matching.

[Galarraga et al., 2014] introduces a hierarchical agglomerative clustering (HAC) approach, where, at any stage, a cluster of descriptions reflects the current belief that all the descriptions in this cluster match. Initially, each description is placed in a cluster of its own. Then, at each iteration, the two most similar clusters are merged, until the similarity of the most similar clusters is below a threshold. The similarity function employed by clustering can be selected from a range of Jaccard-like similarity functions, applied to the values or the attributes of the descriptions. The similarity between two clusters is then calculated using the single-linkage criterion, i.e., the maximum of the pairwise similarities between the contents of each cluster. The complete-linkage, i.e., using the minimum inter-cluster similarity, and the average-linkage criteria were experimentally proven too conservative, leading to many missing matches. The descriptions that belong to the same cluster are finally merged, using a union operation. That is, all the descriptions in the same cluster are assigned the same id. To avoid the cubic, to the number of entity descriptions, time complexity required by HAC, [Galarraga et al., 2014] employs token blocking prior to clustering. Then, HAC is applied within each block, partitioning each block into clusters. Clusters derived from different blocks are merged, if they share a common description.

4.3　RELATIONSHIP-BASED ITERATIVE ENTITY RESOLUTION

In relationship-based iterative ER, the matching decision between two descriptions triggers discovering new candidate pairs for resolution, or re-considering pairs already compared; matched descriptions may be related to other descriptions, which are now more likely to match to each other.

To illustrate the relationships between the descriptions of an entity collection $\mathcal{E}$, usually, an entity graph $G_{\mathcal{E}} = (V, E)$ is used, in which nodes, $V \subseteq \mathcal{E}$, represent entity descriptions and edges, E, reflect the relationships between the nodes. Then, relationship-based ER extends the generic definition of ER (Definition 2.2), by considering the neighborhoods of the descriptions in the match function. For example, such a match function could be of the form:

$$M(e_i, e_j) = \begin{cases} true, & \text{if } sim(nbr(e_i), nbr(e_j)) \geq \theta \\ false, & \text{else,} \end{cases}$$

where sim can be a relational similarity function (like the ones in Section 2.2.2) and θ is a threshold value. Intuitively, the neighborhood $nbr(e)$ of a node e can be the set of nodes that contains e and all the nodes that are connected to e, i.e., $nbr(e) = \{e_j | (e, e_j) \in E\}$, or the set of edges containing e, i.e., $nbr(e) = \{(e, e_j) | (e, e_j) \in E\}$.

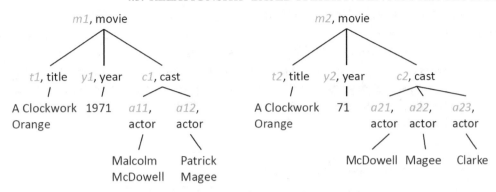

Figure 4.3: Two different descriptions of the movie *A Clockwork Orange* and its cast in XML.

[Weis and Naumann, 2006] study the problem of iterative ER in tree data, and in particular, in XML data. Entity descriptions correspond to XML elements that are composed by text data or other XML elements, and domain experts specify which XML elements are match candidates, thus initializing a priority queue of comparisons. The notion of entity dependency here is used in the following sense: an XML element c depends on another XML element c', if c' is a part of the description of c. Consequently, identifying the matches of c is not independent of identifying the matches of c'. Even if two XML elements are initially considered to be non-matches, they are compared again, if their related elements are found matches. [Weis and Naumann, 2004] uses a similar approach that is based on the intuition that the similarity of two elements reflects the similarity of their data, as well as the similarity of their children. By following a top-down traversal of XML data, the DELPHI containment metric [Ananthakrishna et al., 2002] (see Chapter 2) is used to compare two elements.

Example 4.4   Figure 4.3 shows two different descriptions of the movie *A Clockwork Orange* in XML, represented as a tree. This representation means that the element *movie* consists of the elements *title*, *year*, and *cast*, while the latter further consists of *actor* elements. To identify that the two XML descriptions represent the same movie, we can start by examining the cast of the movies. After we identify that actor a_{11} and actor a_{21} represent the same person, Malcolm McDowell, the chances that the movies m_1 and m_2 match are increased. They are further increased when we find that actors a_{12} and a_{22} also match, representing Patrick Magee. The same matching process over all the sub-elements of m_1 and m_2 will finally lead us to identify that these two descriptions match.

[Bhattacharya and Getoor, 2007] employs an entity graph, following the intuition that two nodes, i.e., entity descriptions, are more likely to match, if their edges, reflecting a relationship between the descriptions, connect to nodes corresponding to the same entity. To capture this inherently iterative intuition, HAC is performed, where, at each iteration, the two most similar (according to one of the relational similarity functions presented in Section 2.2.2) clusters are merged, until the similarity of the most similar clusters is below a threshold. When two clusters are merged, the similarities of their related clusters, i.e., the clusters corresponding to descriptions which are related to the descriptions in the merged cluster, are updated. To avoid the comparison between all the pairs of descriptions when considering the first merge of clusters, a traditional blocking method [McCallum et al., 2000] is employed. This specific approach is known as *collective entity resolution*.

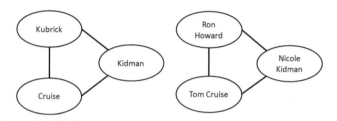

Figure 4.4: An entity graph used by collective entity resolution.

Example 4.5   Figure 4.4 shows an entity graph, in which nodes represent descriptions of persons (actors and directors in this example) and an edge between two persons represents a professional collaboration between the corresponding persons. For example the edge between *Cruise* and *Kidman* represents their co-starring in the movie "Eyes Wide Shut," directed by *Kubrick*, while the edge connecting *Ron Howard* and *Nicole Kidman* represents their partnership for the movie "Far and Away," in which *Tom Cruise* was also starring. *Cruise* and *Tom Cruise* are more likely to represent the same entity, if we know that *Kidman* matches with *Nicole Kidman*, since *Kidman* is a neighbor of *Cruise* and *Nicole Kidman* is a neighbor of *Tom Cruise*.

Using HAC, we first place each description in a distinct cluster and then merge the clusters with the highest similarity. For example, if we decide that the most similar pair of descriptions, based on the values of these descriptions, is that of *Kidman* and *Nicole Kidman*, then we merge the clusters that correspond to these descriptions. Next, we re-calculate the similarity between *Cruise* and *Tom Cruise* and find that this is the next most similar pair, so we also merge this pair of clusters.

[Rastogi et al., 2011] proposes a framework for scaling collective entity resolution to large datasets. This method assumes the existence of a black-box ER algorithm exploiting a set of rules, used as evidence for matching. To achieve scalability, it runs multiple instances of the ER algorithm in small subsets of the entity descriptions (similar to blocking). Since some rules may require the results of more than one block, a message-passing framework is proposed.

In particular, to create the subsets of the descriptions, it uses an extension of blocking, grouping entity descriptions based on not just their similarity, but also on their relational closeness.

For example, it would place the descriptions *Cruise* and *Kubrick* of Figure 4.4 in a common block, not because they are similar, but because they share a collaboration edge. The initial blocks are constructed over the similarity of the descriptions using [McCallum et al., 2000], and then, they are extended taking the *boundary* of each block with respect to entity relationships. The boundary of a block b is defined as the set of descriptions e', for which there is another description e in b, such that e and e' are related. After the construction of such extended blocks, a simple message-passing algorithm is run, to ensure that the match decisions within a block, which might influence the match decisions in other blocks, are propagated to those other blocks. This algorithm retains a list of active blocks, initially containing all blocks. A black-box entity resolution algorithm is run locally, for each active block, and the newly identified matches are added in the result set. Also, all the blocks containing a description of the newly identified matches, are set as active. This iterative algorithm terminates when the list of active blocks becomes empty.

LINDA [Böhm et al., 2012] focuses on identifying matching entity descriptions in entity graphs, constructed from an RDF graph by only considering the (subject, predicate, object) triples, in which subject, predicate, and object are URIs. The evidence of a match between two entity descriptions is based on their common tokens, as well as the identified *sameAs* links between their neighbors in the entity graph. Intuitively, the more matching neighbors that two descriptions have, connected using similar relations, the more likely it is that the two descriptions match. This reasoning proceeds recursively, since matching decisions trigger similarity re-computations between the neighbors of the newly identified matches.

LINDA scales beyond 100 million entities, using MapReduce. Initially, the similarity between all pairs of descriptions is computed, based on $LINDA_{sim}$ (see Chapter 2). Then, the pairs of descriptions are sorted in descending order of similarity and stored in a priority queue. Each machine holds: (i) a partition of this priority queue, defined by a modulo operation on the first description of each pair in the queue, and (ii) the corresponding part of the entity graph, containing the descriptions in the local priority queue partition, along with their neighbors. The iteration step of the algorithm is that, by default, the first pair in the priority queue is considered to be a match and is then removed from the queue and added to the known matches. This knowledge triggers similarity re-computations, which affect the priority queue by enlarging it, when the neighbors of the new match are added again to the queue, re-ordering it, when the neighbors of the identified match move higher in the rank, or shrinking it, by applying transitivity and a unique match per KB constraint. The latter just assumes that an entity description in a KB can only match to one other description in a second KB, i.e., each KB is considered to be clean. The algorithm stops when the priority queue is empty, or when a specific number of iterations has been reached. The only information that is sent on the network is messages about which pairs of descriptions require similarities re-computations.

Example 4.6   Figure 4.5 shows an execution example of the LINDA system for the entity graph shown at the bottom, in which e_3 and e_4 belong to the same KB, while e_1, e_2, and e_5 belong to a second KB. The identified matches are represented by a 1 in the binary symmetric matrix,

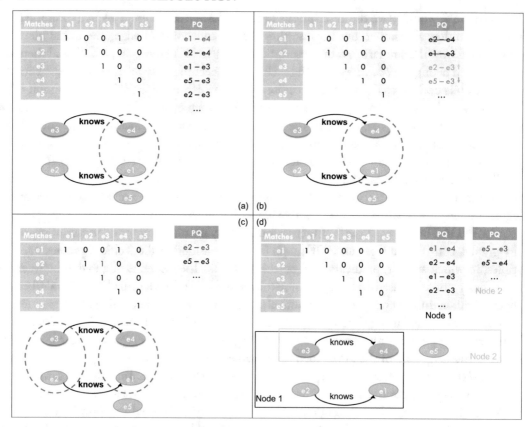

Figure 4.5: An execution example of LINDA. (a) PQ initialization, (b) PQ update, (c) new matches are found, (d) distributed version.

on the top left corners. The entity pair priority queue (depicted as PQ on the top right corners) is initialized and then the top pair (e_1, e_4) is considered a match (Figure 4.5 (a)). This causes the removal of (e_2, e_4) and (e_1, e_3) from PQ, because of the unique match per KB constraint (Figure 4.5 (b)). Since e_1 matches with e_4, it cannot match with any other description from the KB of e_4, and vice versa. The new match also causes a re-ordering of PQ. This happens because the similarity between e_2 and e_3 is increased, since e_2 is a neighbor of e_1 and e_3 is a neighbor of e_4, and they both connect to the matching descriptions with the same predicate ("knows"). So, in Figure 4.5 (c), the top pair, which is by default considered a match, is (e_2, e_3). This causes the removal of the pair (e_5, e_3), again, due to the unique match per KB constraint. After this step, PQ becomes empty and the algorithms terminate, yielding the matches shown in the matrix of Figure 4.5 (c), also illustrated with dashed lines in the entity graph.

An alternative initialization of the algorithm is shown in Figure 4.5 (d), assuming that the algorithm is run on a two-cluster node. The priority queue is divided into two partitions, based on a modulo operation on the first description of the queue. Specifically, pairs starting with e_1 and e_2 are sent to the first node (Node 1), while pairs starting with e_5 are sent to the second node (Node 2). Each node also gets the corresponding partition of the entity graph, containing all the descriptions of its PQ partition, along with their immediate neighbors, as shown at the bottom of Figure 4.5 (d). The same algorithm then runs locally, on each node of the cluster, sharing the knowledge of the identified matches.

[Dong et al., 2005] presents a hybrid approach, based on both partial merging results between descriptions and relations between descriptions, exploiting a graph-based model for iterative ER. In this case, a dependency graph is constructed, in which a node represents the similarity between a pair of entity descriptions and an edge represents the dependency between the matching decisions of two nodes. Hence, if the similarity of a pair of descriptions changes, then we know that the neighbors of this pair might need a similarity re-computation. The dependencies between the matching decisions are distinguished between boolean and real-valued. Boolean dependencies reflect the case in which the similarity of a node only depends on whether the descriptions of its neighbor node match or not, while in real-valued dependencies, the similarity of a node depends on the similarity of the descriptions of its neighbor node. Boolean dependencies are further divided into strong, implying that if a node corresponds to a match, then its neighbor pair should also be a match, and weak, implying that if a node corresponds to a match, then the similarity of its neighbor pair is increased. Initially, all nodes are added to a priority queue. On each iteration, a node is removed from the queue and if the similarity of the node is above a threshold value, its descriptions are merged, aggregating their attribute values, in order to enable further matching decisions. In addition, if the similarity value of this node has increased, its neighbor nodes are added to the priority queue. This iterative process continues until the priority queue becomes empty.

4.4   ITERATIVE BLOCKING

Recent works have proposed using an iterative ER process, interleaved with blocking. Specifically, in iterative blocking, ER is applied to the results of blocking and the results of each iteration potentially alter the initial blocks, triggering a new iteration. Potentially, the block modifications can be either based on relationships between descriptions that have been matched, or on the results of their merging.

[Whang et al., 2009] introduces a merging-based approach for iterative blocking. The intuition here is that the entity resolution results of a processed block may help identify more matches in another block. Unlike other blocking techniques (Chapter 3), it has an additional phase, where newly created entity descriptions in a block (the result of merging the matches) are distributed to other blocks. This process examines one block at a time, looking for matches. When a match is found in a block, the resulting merging of the descriptions that match is propagated to all other blocks, replacing the initial matching descriptions. This way, the comparisons between the same pair of descriptions in different blocks are saved and, in addition, more matches can be identified efficiently. The same block may be processed multiple times, until no new matches are found. This work employs a disk-based algorithm that scales the process of iterative blocking, by fetching data from the disk. Blocks are stored in segments of fixed size, not exceeding the memory size, on the disk. Then, one segment is processed at a time.

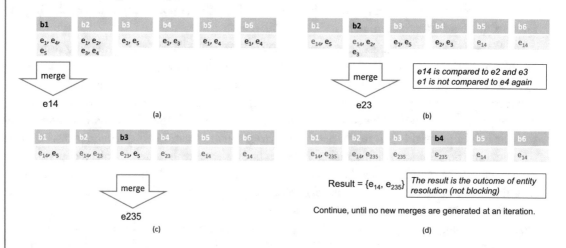

Figure 4.6: An example showing the process of iterative blocking.

Example 4.7 Figure 4.6 shows the process of iterative blocking [Whang et al., 2009], given an initial blocking collection, as the one shown in Figure 4.6 (a). One block is processed at a time, starting from b_1. In this block, the pair (e_1, e_4) is found to be matching. The matching descriptions are merged into a new description e_{14}, representing both. Hence, in the next step (Figure 4.6 (b)), both e_1 and e_4 are replaced, in any block that they appear, by e_{14}. Then, the process moves to the next block b_2, in which a second match is identified, between e_2 and e_3. Similarly, these descriptions are merged as e_{23} and replaced in any block they appear by e_{23} (Figure 4.6 (c)). This iterative process terminates when no new merges are generated in an iteration (Figure 4.6 (d)) and the union $\{e_{14}, e_{235}\}$ of the descriptions is returned as the result of iteratively resolving the initial blocking collection.

HARRA [Kim and Lee, 2010] extends iterative blocking, by employing LSH. Unlike the basic LSH (see Section 2.2.3), which hashes input descriptions to different buckets, corresponding to blocks, in multiple hash tables (one for each band of their minhash signatures), HARRA employs an Iterative Locality-Sensitive Hashing (I-LSH) technique, which reduces space and time requirements, by using a single, re-usable hash table. Specifically, it examines one band at a time, a process illustrated in Figure 4.7. Before placing an entity description in a block, based

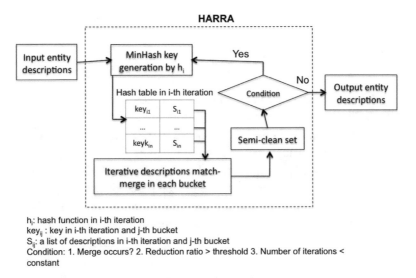

h$_i$: hash function in i-th iteration
key$_{ij}$: key in i-th iteration and j-th bucket
S$_{ij}$: a list of descriptions in i-th iteration and j-th bucket
Condition: 1. Merge occurs? 2. Reduction ratio > threshold 3. Number of iterations <
constant

Figure 4.7: The general structure of HARRA.

on its minhash signature for this band, the description is compared to the contents of the block. If matches are found in this block, then the result of merging these descriptions is placed in the block. Otherwise, the description is placed in the block without any changes. When all descriptions have been scanned, the input is reset with all the descriptions in the hash table, and hashed again until a specific number of iterations has been reached, or the reduction in the number of comparisons is already satisfactory, or no more matches are identified. This way, only one hash table has to be kept in memory on each iteration, avoiding excessive memory requirements for hash tables and candidate pairs of descriptions.

[Malhotra et al., 2014] introduces a parallel, LSH-accelerated iterative blocking technique. The first step is to run a parallel LSH-based blocking via minhashing, with MapReduce. In the next step, a merging-based iterative ER algorithm, like R-Swoosh [Benjelloun et al., 2009], is run on each block, resulting in partially resolved entities, since it is possible for descriptions to be placed into more than one block. This means that the same description may be merged with some descriptions in a block and with some other descriptions in another block. Respecting transitivity, all these descriptions should also be merged. This is achieved by computing the connected components in a graph, in which an edge connects two nodes, i.e., descriptions, if they belong to the same, partially resolved entity. All the connected components are then merged. The latter step is built on the Pregel programming paradigm [Malewicz et al., 2010], implemented by Giraph,[1] which is argued to perform better than MapReduce in such scenaria. An alternative approach for the last step, overcoming the load imbalance between small and large blocks, is to perform, for

[1]http://giraph.apache.org/

each description, the comparisons between itself and all other descriptions with which it shares blocks.

4.5 INCREMENTAL ENTITY RESOLUTION

In this section, we briefly describe some incremental approaches to ER, which avoid recomputing the entire ER workflow, when new entity descriptions are made available. For a detailed overview of the challenges and the optimal solutions proposed in the research literature regarding the problem of incremental ER, the readers are referred to Chapter 3.3 of [Dong and Srivastava, 2015].

[Gruenheid et al., 2014] examines the case of incremental ER when new descriptions are added to an entity collection. In general, the approach proceeds in three steps. First, it performs blocking, and second, for descriptions in the same block, it computes pairwise similarity to construct a graph, in which each node represents a description and each edge between two nodes has a weight reflecting the similarity of the corresponding nodes. Third, it conducts graph clustering, within blocks, such that descriptions of the same entity belong to the same cluster. For resolving the new descriptions in an incremental way, it takes as input only the result of the previous ER process, i.e., the clusters of descriptions within blocks, that are connected in the graph to the newly added descriptions, along with these descriptions. Then, correlation clustering[2] is applied on this set of descriptions, and the new result of ER is achieved after replacing the previous clusters of descriptions with the new ones. It is proved that this method gives optimal results for incremental ER, if optimal methods are employed for correlation clustering. However, correlation clustering is an NP-complete problem, hence, optimal algorithms for it are not feasible for big data; in practice, [Gruenheid et al., 2014] exploits polynomial-time approximation algorithms for correlation clustering that lead to good quality results. Similarly, [Welch et al., 2012] uses the ER results of [Bellare et al., 2013] to incrementally resolve the entities provided as queries in real-time. Again, the entity described in a query is either added to an existing cluster, corresponding to a distinct real-world entity, or creates a new cluster, if it does not match with any other description. [Whang and Garcia-Molina, 2014] further considers that our understanding of data, along with the corresponding matching rules between entity descriptions, evolve frequently. To deal with this aspect, in an incremental manner, it investigates when and how previous materialized ER results can be exploited in order to save work and not re-run ER from scratch.

4.6 PROGRESSIVE ENTITY RESOLUTION

A more recent research direction on ER consists of maximizing the reported matches, given a limited computing budget (e.g., in number of comparisons), by potentially exploiting the partial matching results obtained so far. Such works are called *progressive*[3] and they usually follow the

---

[2]Correlation clustering groups a set of descriptions into the optimum number of clusters without specifying that number in advance.

[3]According to [Whang et al., 2013b], at any time, a progressive ER algorithm yields more matches than a non-progressive one, while both yield the same final results.

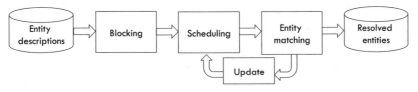

Figure 4.8: Progressive ER process.

workflow of Figure 4.8. They essentially extend the typical ER workflow (see Chapter 2.1) with a scheduling phase, which is responsible for selecting which pairs of descriptions, that have resulted from *blocking*, will be compared in the *entity matching* phase and in what order. The goal of this new phase is to favor more promising comparisons, i.e., those that are more likely to result in matches. This way, those comparisons are executed before less promising ones and thus, more matches are identified early on in the process. Moreover, the optional *update* phase propagates the results of matching, such that a new scheduling phase will promote the comparison of pairs that were influenced by the previous matches. This iterative process continues until the pre-defined computing budget is consumed.

[Altowim et al., 2014] exploits the dependency graph of [Dong et al., 2005] (in which nodes are description pairs, i.e., candidate matches, and an edge indicates that the resolution of a node influences the resolution of another node) in a progressive ER setting. A black-box blocking phase is used, to avoid building a dependency graph with all the description pairs. Then, for the scheduling phase, this approach divides the total cost budget into several windows of equal cost. For each window, a comparison schedule is generated, by choosing among the candidate schedules, i.e., those whose cost does not exceed the current window, the one with the highest expected benefit. The cost of a schedule is computed by considering the cost of finding the description pairs in a block according to the available storage policy (in memory/disk/cloud), and the cost of resolving every description pair.[4] The benefit of a schedule is determined by how many matches are expected to be found by this schedule (direct benefit), and how useful it will be to declare those nodes as matches, in identifying more matches within the cost budget (indirect benefit). In general, a node is more likely to be a match, when it is influenced by more matching nodes, and it is more influential, when it is expected to be a match and it has many direct dependent nodes. In the update phase, after a schedule is executed, the matching decisions are propagated to all the influenced nodes, whose expected benefit now increases and have, thus, higher chances of being chosen by the next schedule. The algorithm terminates when the cost budget has been reached, and all the unresolved pairs are considered non-matches, since, statistically, matches are significantly fewer than non-matches.

---

[4]The matching decision is based on the application of several similarity functions, one for each of the attributes of a description with a fixed schema.

[Whang et al., 2013b] proposes three different heuristics for the scheduling phase of a progressive ER workflow, assuming a static execution (i.e., no update phase). The first heuristic relies on a sorted list of description pairs, which can be generated by a blocking key as in sorted neighborhood (see Chapter 3.2), and then iteratively considers sliding windows of increasing size, comparing descriptions of increasing distance. Starting from a window of size 2, this heuristic favors comparisons of descriptions with more similar values on their blocking keys. The second heuristic is a hierarchy of description partitions. Such a hierarchy is built by applying different distance thresholds for each partitioning: highly similar descriptions are placed in the lower levels, while somehow similar descriptions in higher ones. By traversing bottom-up this hierarchy until the cost budget has been reached, this scheduling policy favors the resolution of highly similar descriptions. Given a partition hierarchy, the third heuristic creates a sorted list of descriptions (and not pairs of descriptions), placing first in the list, descriptions that appear in bigger partitions in this hierarchy. A description that appears in big partitions is expected to match with many other descriptions. Hence, by placing the descriptions of the biggest partitions first in the list, many matches are expected to be identified in the first few comparisons performed by the ER algorithm, when resolved sequentially. However, the success of this heuristic depends on the ER algorithm that will be used. This work also introduces two different early termination policies: in the first policy, the algorithm terminates when the rate of newly found matches drops below a specific threshold, while in the second policy, the algorithm stops when a given percentage of the total comparisons has been performed.

More recently, [Papenbrock et al., 2015] introduces a progressive sorted neighborhood method extending the first heuristic of [Whang et al., 2013b] (sorted list of description pairs), to capture cases in which matches appear in dense areas of the initial sorting, by adding a scheduling phase that performs a local lookahead—if the descriptions at positions (i, j) are found to match, then the descriptions at $(i + 1, j)$ and $(i, j + 1)$ are immediately compared, since they have a high chance of matching. The lookahead step is performed recursively also for the newly found matches of a previous lookahead, before sliding the window, thus preferring locally promising comparisons in the otherwise static execution. Another variation of sorted neighborhood is proposed, which partitions the sorted descriptions into blocks of the same size (i.e., non-overlapping windows) and initially compares the descriptions within each block. Then, in each scheduling phase, it selects the blocks that delivered the most matches in the previous iteration and compares the contents of those blocks, with the contents of their neighbor blocks (blocks follow the ranking of the descriptions). In case of ties, block pairs with a smaller rank-distance are selected, since the distance in the sort rank still defines the expected similarity of the descriptions.

As a final note, we can also consider meta-blocking [Papadakis et al., 2014a] (presented in Chapter 3) as a progressive ER algorithm, whose scheduling heuristic is to retain the top-k edges of the blocking graph. The edges of the blocking graph (corresponding to pairs of descriptions), can be sorted in descending order of their weights (corresponding to the likelihood that they match), while the whole ER computation should not exceed k comparisons. This algorithm

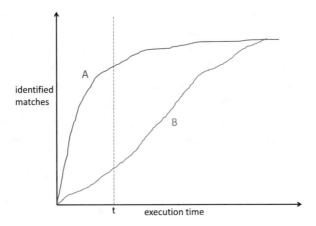

Figure 4.9: A progressive ER algorithm A, compared to a typical ER algorithm B.

follows a static execution, since there is no update phase, i.e., the sorting of edges occurs only once and the previous matching decisions do not affect the following ones.

4.7 DISCUSSION

The inherent distribution of entity descriptions in different KBs along with their significant semantic and structural diversity yield incomplete evidence regarding the similarity of descriptions published in the Web of data. Hence, typical ER approaches consisting of distinct blocking and matching phases may result in a significant number of missed matches especially between central and peripheral KBs in the LOD cloud (see Chapter 5). Iterative ER approaches aim to tackle this problem by exploiting any partial result of the ER process in order to generate new candidate pairs of descriptions not considered in a previous step, by eventually re-considering the same candidate pairs of descriptions (something that is prohibiting in the Web scale). Moving forward, recent works have suggested iteratively generating the results of an ER process in a progressive, pay-as-you-go fashion. Their main goal is not to identify all possible matches, but to dynamically plan an ER strategy that will lead to as many matches as possible, given a budget constraint. Figure 4.9 illustrates the underlying: a progressive ER algorithm A, at any specific time point t, has identified more matches than a typical ER algorithm B, while, ideally, they have both identified the same matches eventually. So far, only the work of[Altowim et al., 2014] can handle entity descriptions of multiple types and exploit their structural relationships, while all other progressive algorithms concern flat descriptions of a single entity type. A detailed characterization of the quality of obtained matches (e.g., by maximizing either the number or the size of generated partitions with resolved descriptions) achieved by progressive ER algorithms in the Web of data is an open issue.

Existing iterative approaches that are based on merges, typically exploit content-based similarity functions (see Section 2.2.1), applied on the merge result of matching descriptions, since the content of a merge result is different from the content of the individual descriptions that compose it. Merging-based iterative ER is mostly suitable for deduplication, i.e., a dirty ER task, since the merge operation may not be applicable to the descriptions of autonomous KBs in the Web of data. On the other hand, iterative approaches that are based on entity relationships, typically exploit relational similarity functions, like the ones discussed in Section 2.2.2. Specifically, these approaches revise their matching decisions, based on the propagation of similarity from neighborhood descriptions that have been identified to match. Finally, existing iterative blocking algorithms, interleaving entity resolution with blocking, are based on merges. Hence, they are also more suited for a dirty ER task. Inherently, they follow a sequential execution model, since the results of ER in one block directly affect other blocks.

CHAPTER 5

Experimental Evaluation of Blocking Algorithms

In this chapter, we present the experimental framework we have designed for a critical assessment of blocking algorithms. In particular, we describe the datasets and the measures we employed to study the behavior of the blocking algorithms under different semantic and structural characteristics of entity descriptions in the Linked Open Data (LOD) cloud. Then, we analyze the quality and performance of the evaluated blocking techniques, taking into consideration the specific features of each dataset. Finally, we present the results of blocking when different kinds of links, other than *owl:sameAs*, are used as ground truth and conclude with a discussion of the lessons learned from this analysis.

In our evaluation, we use the adaptation of token blocking ToB, attribute clustering blocking AtC and prefix-infix(-suffix) blocking PIS in MapReduce. All these blocking techniques are based on the notion of common tokens, which is the core of token-based similarity functions (see Section 2.2). Both ToB and PIS can be used with either *clean-clean Cl* or *dirty Di* entity collections, while AtC is suitable for Cl collections. Moreover, the process of AtC requires a similarity function; we use Jaccard similarity over the set of trigrams from the values (similar to [Papadakis et al., 2013]). For our experiments, we have used a cluster of 16 Ubuntu 12.04.3 LTS servers (virtual machines), each with 8 CPUs, 8GB RAM, and 60GB of hard disk capacity, placed in a local network with network speed around 500 MB/s. One of the nodes served only as a master node and the rest were slave nodes. Each node could run simultaneously four map or reduce tasks, while each task had a heap size of 1250 MB available. Finally, we used the *sim*okeanos[1] GRNET cloud service, Apache Hadoop 1.2.0 and Java version 1.7.0_25 from OpenJDK.

5.1 DATASETS

Our empirical study relies on real data from the Billion Triples Challenge dataset of 2012[2] (BTC12), DBpedia 3.5, the Kasabi data publishing platform,[3] and the Linked Archives Hub project.[4]

[1]okeanos.grnet.gr
[2]km.aifb.kit.edu/projects/btc-2012/
[3]archive.org/details/kasabi
[4]data.archiveshub.ac.uk/

To capture the differences in the heterogeneity and semantic relationships of entity descriptions, we distinguish between data originating from sources in the *center* and the *periphery* of the LOD cloud. In general, central KBs, such as DBpedia and Freebase are derived from a common source, Wikipedia, from which they extract information regarding an entity. Such entity descriptions often refer to the original wiki page and they feature synonym attributes whose values share a significant number of common tokens. Since they have been exhaustively studied in the literature, entity descriptions across central LOD KBs are heavily interlinked using in their majority *owl:sameAs* links.[5] In our experiments, we used DBpedia (version 3.7) and Freebase, both coming from BTC12 (*BTC12DBpedia* and *BTC12Freebase*, respectively), as well as the raw infoboxes dataset from DBpedia 3.5 (*Infoboxes*), i.e., essentially two different versions of DBpedia. We additionally included a movies dataset,[6] used in [Papadakis et al., 2013], extracted from DBpedia movies and *IMDB*, in order to verify the results of this work and validate the correctness of our algorithms.

On the other hand, KBs in the periphery of the LOD cloud are far more thematically diverse and thus, sparsely interlinked with central KBs. Due to their cross-domain nature, we believe that they can benefit from entity resolution and blocking algorithms in the Web of data. In our experiments, we considered the *BTC12Rest*, the *BBCmusic*, and the *LOCAH* datasets. *BTC12Rest* originates from the BTC12 dataset, which, in turn, consists of multiple KBs like DBLP (with bibliographic data), geonames and drugbank. *BBCmusic* originates from Kasabi and contains descriptions regarding music bands and artists, which are extracted from MusicBrainz and Wikipedia. For *LOCAH*, we used the latest published version at Archives hub (on March 4th, 2014). This, rather small dataset links descriptions of people, from UK archival institutions, with their corresponding descriptions in DBpedia.

Table 5.1 provides detailed statistics about these datasets, regarding the number of contained triples, entity descriptions, attributes, as well as the average number of attribute-value pairs per description. We have also included the number of distinct entity types, taken as the distinct values of the property *rdf:type*, when provided. Since in *Infoboxes*, this property is not available, we used the values of the *wikiPageUsesTemplate* instead. It is interesting to observe that *BTC12DBpedia* contains more types than attributes. This is due to the fact that DBpedia entities may have multiple types from taxonomic ontologies like Yago. Note also that *IMDB* is the dataset with the highest number of attribute-value pairs per description. We have finally included in each dataset the number of duplicate entity descriptions based on our ground truth, i.e., entity descriptions that have been reported to be equivalent (via *owl:sameAs* links) across all datasets of our testbed. Taking into account the transitivity of equality, those descriptions should be regarded as matches, too.

To investigate the ability of existing blocking algorithms in recognizing relatedness links beyond the *owl:sameAs* among entity descriptions, we finally considered the Kasabi *airports* and

---

[5]DBpedia is the one with the highest in-degree while 80% of cross source links in LOD are *owl:sameAs* [Schmachtenberg et al., 2014].
[6]13s.de/ papadakis/erFramework.html

Table 5.1: Datasets characteristics

	BTC12DBpedia	Infoboxes	BTC12Rest	BTC12Freebase	BBCmusic	LOCAH	DBpedia$_{mov}$	IMDB
RDF triples	102,306,242	27,011,880	849,656	25,050,970	268,759	12,932	180,680	816,012
entity descriptions	8,945,920	1,638,149	31,668	1,849,180	25,359	1,233	27,615	23,182
avg. attribute-value pairs per description	11.44	16.49	26.83	13.55	10.60	10.49	6.54	35.20
attributes	36,354	31,857	518	8,323	29	14	5	7
entity types	258,202	5,535	33	8,232	4	4	1	1
attributes/entity types	0.14	5.76	15.7	1.01	7.25	3.5	5	7
duplicates	0	0	863	12,058	372	250	0	0

Table 5.2: Characteristics of datasets with different types of links to *BTC12DBpedia*

	airports	airlines	twitter	books	iati		www2012
RDF triples	238,973	15,465	6,743	2,993	378,130		11,772
entity descriptions	12,294	1,141	2,932	748	31,868		1,547
avg. attribute-value pairs per description	19.44	13.55	2.30	4.00	11.87		7.61
link	umbel:isLike	umbel:isLike	dct:subject	dct:subject	dct:subject	dct:coverage	foaf:based_near
links	12,269	1,217	20,671	1,605	23,763	7,833	1,562

airlines datasets, containing data linked to DBpedia, which is the dataset with the highest number of references, with the *umbel:isLike* property. This property is used to associate entities that may or may not be equivalent, but are believed to be so. The *twitter* dataset contains data for the presentations of an ESWC conference. It is linked to DBpedia with the *dct:subject* property, which captures relatedness of entities to topics and it is also used in the *books* and *iati* datasets. *Books* describes books listed in the English language section of Dutch printed book auction catalogues of collections of scholars and religious ministers from the 17th century. *Iati* contains data from the International Aid Transparency Initiative. *Iati* is also connected to DBpedia with the *dct:coverage* property, which associates an entity to its spatial or temporal topic, its spatial applicability, or the jurisdiction under which it is relevant. Finally, the *www2012* dataset contains data from the WWW2012 conference, linked to DBpedia with the *foaf:based_near* property, which associates an entity to an abstract notion of location. Table 5.2 details the type and the number of links of these datasets to DBpedia.

In this setting, we combine *BTC12DBpedia* with each of the datasets of Table 5.1 to produce the entity collections presented in Table 5.3, on which we finally ran our experiments.

- *D*1 combines *BTC12DBpedia* with *Infoboxes*. Since it contains two versions of the same dataset, it is considered as a homogeneous collection. This is the biggest collection in terms of triples, as well as attributes. It also contains the highest percentage of matches with matching neighbors (from a sample of 1,000 matches), i.e., entity descriptions that are either identical or known matches.

- *D*2 combines *BTC12DBpedia* with *BTC12Rest*. Since it is constructed by many different datasets, it is the most heterogeneous collection. Note that *BTC12Rest* has the highest number of attributes per entity type.

Table 5.3: Entity collections characteristics

	D1	D2	D3	D4	D5	D6
RDF triples	129,318,122	103,155,898	127,357,212	102,575,001	102,319,174	996,692
entity descriptions	10,584,069	8,977,588	10,795,100	8,971,279	8,947,153	50,797
avg attribute-value pairs per description	12.22	11.49	11.80	11.43	11.44	19.62
attributes	68,211	36,872	44,677	36,383	36,368	12
entity types	263,737	258,232	266,434	258,206	258,205	1
matches	1,564,311	30,864	1,688,606	23,572	1,087	22,405
matches (incl. duplicates)	1,564,311	31,727	1,700,664	23,944	1,337	22,405
matches/non-matches	$1.07 \cdot 10^{-7}$	$1.09 \cdot 10^{-7}$	$1.02 \cdot 10^{-7}$	$1.04 \cdot 10^{-7}$	$9.85 \cdot 10^{-8}$	$3.5 \cdot 10^{-5}$
matches/non-matches (dirty)	$2.79 \cdot 10^{-8}$	$7.87 \cdot 10^{-10}$	$2.92 \cdot 10^{-8}$	$5.95 \cdot 10^{-10}$	$3.34 \cdot 10^{-11}$	$1.74 \cdot 10^{-5}$
%matches with matching neighbors (sampled)	86.3%	25.5%	0.007%	30.1%	59.8%	0%
comparisons (w/o blocking)						
clean–clean	$1.47 \cdot 10^{13}$	$2.83 \cdot 10^{11}$	$1.65 \cdot 10^{13}$	$2.27 \cdot 10^{11}$	$1.1 \cdot 10^{10}$	$6.4 \cdot 10^{8}$
dirty	$5.6 \cdot 10^{13}$	$4.03 \cdot 10^{13}$	$5.83 \cdot 10^{13}$	$4.02 \cdot 10^{13}$	$4 \cdot 10^{13}$	$1.29 \cdot 10^{9}$

- *D*3 combines *BTC12DBpedia* with *BTC12Freebase*. It is the biggest collection in terms of entity descriptions, matches, entity types, and comparisons.

- *D*4 combines *BTC12DBpedia* with *BBCmusic*. It is the collection with the lowest number of attribute-value pairs per entity description.

- *D*5 combines *BTC12DBpedia* with *LOCAH*. *LOCAH* is the smallest dataset, both in terms of triples and entity descriptions.

- *D*6 combines DBpedia movies and *IMDB*, as originally used in [Papadakis et al., 2013]. It is the most homogeneous collection, since both of its datasets contain descriptions only of movies (i.e., a single entity type) using the smallest number of attributes among all collections. However, the significantly greater (even by six orders of magnitude, compared to the other collections) ratio of matches to non-matches is not typical of the collections we can find in the Web of data.

Following the distinction of our datasets between central and peripheral, we also distinguish our collections between central (*D*1, *D*3, and *D*6), composed of only central datasets and peripheral (*D*2, *D*4, and *D*5), part of which are peripheral datasets. For all these collections, we consider both their *clean-clean* and *dirty* versions. In practice, for our datasets, the *clean-clean* and *dirty* versions of a collection are the same; their distinction serves only as a means for measuring how well a blocking technique can identify links across different datasets and within the same dataset. We finally combine *BTC12DBpedia* with each peripheral dataset of Table 5.2 to produce entity collections for studying the ability of blocking algorithms to discover different relatedness attributes.

Ground Truth. As a ground truth of matching entity descriptions for *D*2–*D*5, we consider the *owl:sameAs* links to/from DBpedia 3.7 (the version used in BTC12). As a ground truth for *D*1, we consider the subject URIs of *Infoboxes* that also appear as subjects in *BTC12DBpedia*.

Table 5.4: Definitions for pairs of descriptions, based on whether they appear in a common block, or not

		Ground-truth	
		match	non-match
Blocking result	candidate match	TP	FP
	not a candidate match	FN	TN

The ground truth of $D6$ is made of DBpedia movies that are connected with IMDB movies through the *imdbId* propert.[7] Based on the ground truth and the generated blocks, we label a pair of entity descriptions according to Table 5.4. Intuitively, we say that a known matching pair of descriptions is correctly resolved, i.e., a true positive (TP), if there is at least a block, to which both these descriptions belong. Pairs of descriptions belonging to the same block are candidate matches. A false positive (FP) is a distinct candidate match that is not contained in the ground truth. Conversely, if a known matching pair is not a candidate match, then this pair is considered as a false negative (FN). All other pairs of descriptions are considered to be true negatives (TN).

Similarly to $D2$–$D5$, we used the available types of links of the datasets of Table 5.2 to *BTC12- DBpedia*, instead of *owl:sameAs*, to produce the ground truth of the corresponding entity collections. From all datasets, except $D6$, we removed the triples that are present in the ground truth, since identifying those links is the goal of our tasks.

Pre-processing. We used a pre-processing program, implemented in MapReduce that parses RDF triples in order to transform them into entity descriptions, which are the input of the methods used in our study. It simply groups the triples by subject, and outputs each group as an entity description, using the subject as the entity identifier. We kept only the entity descriptions for which we know their linked description in *BTC12DBpedia* and removed the rest. This way, we know that any suggested comparison between a pair of descriptions outside the ground truth is false. Moreover, we removed triples containing a blank node.[8] Last but not least, we applied tokenization, lowercasing, and removed non-English literals (when such information was explicitly provided by a language tag), as a typical standardization step of the literal values (e.g., similar to [Isele and Bizer, 2012, 2013]).

5.2   MEASURES

The employed quality measures along with a short description are summarized in Table 5.5. The ideal value of each measure is in boldface. Recall shows how many of the matching descriptions, the blocking method manages to place together into at least one common block. The recall of a blocking technique is the upper recall threshold of a non-iterative entity resolution algorithm, which takes its generated blocks as input. Therefore, recall represents the effectiveness of block-

---

[7]l3s.de/ papadakis/erFramework.html

[8]Anonymous or blank nodes, that play the role of existential variables, appearing in subjects and objects, should be avoided when datasets are published according to the Linked Data paradigm.

Table 5.5: Quality measures (the ideal value of each measure is in boldface)

Name	Formula	Range	Description
Recall	$\frac{TP}{TP+FN}$	$[0, \mathbf{1}]$	Measure what fraction of the known matches are candidate matches.
Precision	$\frac{TP}{TP+FP}$	$[0, \mathbf{1}]$	Measure what fraction of the candidate matches are known matches.
F-measure	$2\frac{Precision \cdot Recall}{Precision + Recall}$	$[0, \mathbf{1}]$	The harmonic mean of precision and recall.
RR	$1 - \frac{\text{comparisons with blocking}}{\text{comparisons without blocking}}$	$[0, \mathbf{1}]$	Returns the ratio of reduced comparisons when blocking is applied.
H3R	$2\frac{RR \cdot Recall}{RR + Recall}$	$[0, \mathbf{1}]$	The harmonic mean of recall and reduction ratio.
$H3R_w$	$(1 + w^2)\frac{RR \cdot Recall}{(w^2 \cdot RR) + Recall}$	$[0, \mathbf{1}]$	The weighted harmonic mean of recall and reduction ratio, with recall having w times the weight of reduction ratio.

ing. Seen differently, (1-recall) represents the *cost of blocking*. Precision shows how many of the suggested comparisons will be redundant, i.e., between non-matches. Reduction ratio, RR, is the percentage of comparisons that we save if we apply the given blocking method. Consequently, it reflects the *benefit of blocking*, since the reason for using blocking in the first place is the reduction in the required comparisons.

In general, a good blocking method should have a low impact on recall, i.e., a low cost, and at the same time, a great impact on the number of required comparisons, i.e., a high benefit. Typically, this trade-off is measured by the F-measure, namely, the harmonic mean of recall and precision. However, as we will see in the next section, the values of F-measure are dominated by the values of precision, which are many orders of magnitude lower than those of recall, so the F-measure cannot be easily used to express this trade-off. Instead, we introduce $H3R$ as the harmonic mean of recall and reduction ratio. Similar to the F-measure, $H3R$ gives high values only when both recall and reduction ratio have high values. Unlike F-measure, $H3R$ manages to capture the trade-off between effectiveness and efficiency in a more balanced way. In case emphasis is put on recall, i.e., minimizing the cost of blocking, we can also use a weighted mean, $H3R_w$, which considers recall to be w times more important than reduction ratio. In our experiments, for instance, we have also included the scores for $H3R_2$. Whether more emphasis should be put on recall or reduction ratio, depends on the dataset characteristics. On a homogeneous entity collection, where a high recall is not difficult to achieve, more weight can be given to reduction ratio. On the other hand, on a heterogeneous collection, like many of the ones met in the Web of data, getting as many matches as possible is usually the first priority, making recall more important than reduction ratio. As a final notice, observe that $H3R$ does not estimate the performance of a blocking approach (as, for example, [Papadakis et al., 2012] does), but evaluates it based on the actual results.

5.3   QUALITY RESULTS

5.3.1   IDENTIFIED MATCHES (TPS)

Token blocking: The basic premise of this algorithm is that matching entity descriptions should at least share a common token, disregarding the comparisons between descriptions that do not share any common tokens. Therefore, the higher the number of common tokens, i.e., tokens shared by the datasets composing an entity collection, a description has, the higher the chances it will be placed in a block with a matching description, increasing recall. Figure 5.1 (top) presents the distributions of common tokens per entity description, showing that descriptions in central collections feature many more common tokens than those in peripheral collections.[9] For example, 41.43% and 44% of descriptions in $D1$ and $D3$, respectively, have two to four common tokens, while for $D2$, $D4$, and $D5$ the corresponding values are 33.26%, 26.03%, and 12.97%, respectively. We observe a big difference in the distribution of $D6$, which contains many more common tokens per description, to those of the other collections, due to the fact that the ratio of matches to non-matches is much higher than in the other collections (Table 5.3). Only 23.75% of the descriptions in this collection have zero to ten common tokens. This figure also shows that a big number of descriptions in peripheral collections do not share any common tokens. Those are hints that the recall of token blocking in central collections is higher than in peripheral collections.

Indeed, $D6$ is the dataset with the highest recall (99.92%) and the highest number of common tokens per entity (19), while $D5$ is the dataset with the lowest recall (72.13%) and number of common tokens per entity (0). There is a big difference in the number of common tokens in $D6$, compared to $D1$ and $D3$, which is not reflected by their small difference in recall. Due to the high ratio of matches to non-matches in $D6$ (Table 5.3), descriptions in this collection have many common tokens and this leads to high recall.

Attribute clustering blocking: The goal of attribute clustering blocking is to improve the precision of token blocking, while retaining its recall as much as possible (it cannot have higher recall). To do this, it restricts the number of attributes on which entity descriptions, featuring a common token, should be compared. Comparisons between descriptions that do not share a common token in a common attribute cluster are thus discarded. Hence, descriptions with many common tokens in common clusters are more likely to be matched. Figure 5.1 (bottom) presents the distributions of the number of common tokens in common attribute clusters per entity description. It shows a clearer distinction between central and peripheral collections than Figure 5.1 (top); the descriptions in central collections have many more common tokens in common clusters, while many descriptions in peripheral collections do not have any common token in a common cluster. This occurs because values in the descriptions of peripheral collections are much less similar than those of central collections, leading to a bad clustering of the attributes and, consequently, to lower recall. In fact, $D6$ is the dataset with the highest recall (99.55%) and the highest number of common tokens in common attribute clusters per description (19). On the other hand, $D2$ and

---

[9]We take the median values and not the averages, as the latter are highly influenced by extreme values and our distributions are skewed.

$D5$, which have the lowest recall values (68.42% and 71.11%, respectively) also have the lowest number of common tokens in common attribute clusters per description (0).

In central collections ($D1$, $D3$, $D6$), many small clusters of similar attributes are formed, as the values of the descriptions are also similar. This leads to a minor decrease in recall, compared to token blocking, while it significantly improves its precision (even by an order of magnitude in $D3$). $D1$ forms many (16,886), small attribute clusters (of two attributes in the median case), since in most cases there is a one-to-one mapping between the attributes of the datasets that compose it. These clusters correspond to the mapping of the same attribute used by the two versions of DBpedia that compose this collection.

However, this approach has a substantial impact on recall in peripheral collections ($D2$, $D4$, $D5$), even if it still improves precision in all collections (even by an order of magnitude for $D4$). The descriptions in those collections have only a few common tokens in the first place, which leads to a bad clustering of attributes; few clusters of many attributes, not similar to each other, are formed. Hence, if we make the blocking criterion of token blocking stricter, by also considering attributes, then the more distinct attributes used per entity type, the more difficult it is for an entity description, to be placed in a common block with a matching description. For *BTC12Rest* (part of $D2$), the ratio between attributes and entity types (last row of Table 5.1) is the highest (15.7), leading to a great impact on recall (-24.04%). This dataset has the biggest number of data sources that compose it and many different attribute names can be used for the same purpose; hence, big attribute clusters are formed. *LOCAH* (part of $D5$) only has 3.5 attributes per entity type. Thus, the recall of attribute clustering blocking is insignificantly reduced (-1.02%), compared to that of token blocking.

Prefix-Infix(-Suffix) blocking: Prefix-Infix(-Suffix) blocking is built on the premise that many URIs contain useful information. Its goal is to extend token blocking and improve both its recall, by also considering the subject URIs of the descriptions, and its precision, by disregarding some unneeded tokens in the URI values (either in the prefix or suffix). It achieves good recall values in datasets with similar naming policies in the URIs, as in $D4$, part of which is *BBCmusic*, which also has Wikipedia as a source. However, it misses many matching pairs of descriptions, when the names of the URIs do not contain useful information, as in $D3$ that uses random strings as ids, or have different policies, as in $D5$, which uses concatenations of tokens, without delimiters, as URIs. The recall of $D1$ is 100%, because the collection is constructed this way; it consists of two versions of the same dataset, DBpedia, and the URIs appearing as subjects in *Infoboxes* are only those URIs that also appear as subjects in *BTC12DBpedia*. PIS is not applicable (marked as N/A) to $D6$, since its URIs have been replaced with numerical identifiers in the provided datasets.[10]

---

[10]We want to keep this dataset unchanged, as its only purpose is to verify the correctness of our results.

Table 5.6: Statistics and evaluation of blocking methods

	D1	D2	D3	D4	D5	D6
Token blocking statistics:						
blocks	1,639,962	122,340	1,019,501	57,085	2,109	40,304
comparisons (clean-clean)	$1.68 \cdot 10^{12}$	$3.74 \cdot 10^{10}$	$6.56 \cdot 10^{11}$	$2.39 \cdot 10^{10}$	$8.72 \cdot 10^{8}$	$2.91 \cdot 10^{8}$
RR (clean)	88.51%	86.81%	96.03%	89.48%	92.09%	54.50%
comparisons (dirty)	$5.56 \cdot 10^{12}$	$3.68 \cdot 10^{12}$	$4.27 \cdot 10^{12}$	$4.02 \cdot 10^{12}$	$1.01 \cdot 10^{12}$	$2.05 \cdot 10^{9}$
RR (dirty)	90.08%	90.87%	92.67%	90.01%	97.48%	−58.85%
common tokens per entity	4	3	4	2	0	19
Attribute clustering blocking statistics:						
blocks	5,602,644	150,293	1,673,855	39,587	3,724	43,716
comparisons	$3.22 \cdot 10^{11}$	$4.20 \cdot 10^{9}$	$1.84 \cdot 10^{11}$	$1.43 \cdot 10^{9}$	$7.13 \cdot 10^{8}$	$2.13 \cdot 10^{8}$
RR	97.80%	98.52%	98.89%	99.37%	93.54%	66.80%
common tokens in common attribute clusters per entity	4	0	4	2	0	19
attribute clusters	16,886	124	2,106	6	8	4
attributes per attribute cluster	2	142	9	4,261	3,946	3
Prefix-Infix(-Suffix) blocking statistics:						
blocks	3,266,798	141,517	789,723	45,403	2,098	N/A
comparisons (clean-clean)	$1.10 \cdot 10^{12}$	$1.78 \cdot 10^{10}$	$2.75 \cdot 10^{11}$	$2.30 \cdot 10^{9}$	$4.08 \cdot 10^{8}$	N/A
RR (clean)	92.48%	93.72%	98.34%	98.99%	96.30%	N/A
comparisons (dirty)	$4.39 \cdot 10^{12}$	$3.45 \cdot 10^{12}$	$5.34 \cdot 10^{12}$	$3.32 \cdot 10^{12}$	$1.76 \cdot 10^{12}$	N/A
RR (dirty)	92.16%	91.44%	90.84%	91.76%	95.59%	N/A
Recall:						
Token blocking (clean-clean)	98.38%	92.46%	95.52%	87.76%	72.13%	99.92%
Token blocking (dirty)	98.38%	89.99%	94.85%	87.95%	77.34%	99.92%
Attribute clustering blocking	97.31%	68.42%	92.10%	76.84%	71.11%	99.55%
Prefix-Infix(-Suffix) blocking (clean-clean)	100%	91.71%	87.68%	95.44%	68.17%	N/A
Prefix-Infix(-Suffix) blocking (dirty)	100%	89.25%	87.06%	95.50%	74.12%	N/A
Precision:						
Token blocking (clean-clean)	$1.56 \cdot 10^{-6}$	$1.00 \cdot 10^{-6}$	$2.49 \cdot 10^{-6}$	$1.30 \cdot 10^{-6}$	$1.13 \cdot 10^{-6}$	$1.21 \cdot 10^{-4}$
Token blocking (dirty)	$3.64 \cdot 10^{-7}$	$5.14 \cdot 10^{-9}$	$3.78 \cdot 10^{-7}$	$1.05 \cdot 10^{-8}$	$1.29 \cdot 10^{-9}$	$7.51 \cdot 10^{-5}$
Attribute clustering blocking	$8.51 \cdot 10^{-6}$	$5.76 \cdot 10^{-6}$	$1.01 \cdot 10^{-5}$	$1.41 \cdot 10^{-5}$	$1.35 \cdot 10^{-6}$	$1.52 \cdot 10^{-4}$
Prefix-Infix(-Suffix) blocking (clean-clean)	$1.87 \cdot 10^{-6}$	$2.19 \cdot 10^{-6}$	$5.72 \cdot 10^{-6}$	$1.01 \cdot 10^{-5}$	$2.05 \cdot 10^{-6}$	N/A
Prefix-Infix(-Suffix) blocking (dirty)	$6.04 \cdot 10^{-7}$	$8.21 \cdot 10^{-9}$	$2.77 \cdot 10^{-7}$	$1.23 \cdot 10^{-8}$	$6.99 \cdot 10^{-10}$	N/A
F-measure:						
Token blocking (clean-clean)	$3.13 \cdot 10^{-6}$	$2.00 \cdot 10^{-6}$	$9.72 \cdot 10^{-7}$	$2.06 \cdot 10^{-8}$	$1.94 \cdot 10^{-9}$	$2.42 \cdot 10^{-4}$
Token blocking (dirty)	$7.28 \cdot 10^{-7}$	$1.03 \cdot 10^{-8}$	$7.55 \cdot 10^{-7}$	$2.10 \cdot 10^{-8}$	$2.59 \cdot 10^{-9}$	$1.50 \cdot 10^{-4}$
Attribute clustering blocking	$1.70 \cdot 10^{-5}$	$1.15 \cdot 10^{-5}$	$2.02 \cdot 10^{-5}$	$2.82 \cdot 10^{-5}$	$2.69 \cdot 10^{-6}$	$3.04 \cdot 10^{-4}$
Prefix-Infix(-Suffix) blocking (clean-clean)	$3.75 \cdot 10^{-6}$	$4.38 \cdot 10^{-6}$	$9.98 \cdot 10^{-7}$	$2.02 \cdot 10^{-5}$	$4.11 \cdot 10^{-6}$	N/A
Prefix-Infix(-Suffix) blocking (dirty)	$1.21 \cdot 10^{-6}$	$1.64 \cdot 10^{-8}$	$5.55 \cdot 10^{-7}$	$2.46 \cdot 10^{-8}$	$1.40 \cdot 10^{-9}$	N/A
$H3R$:						
Token blocking (clean-clean)	93.18%	89.55%	95.77%	88.61%	80.90%	70.53%
Token blocking (dirty)	94.05%	90.43%	93.75%	88.97%	86.25%	N/A ($RR < 0$)
Attribute clustering blocking	97.55%	80.76%	95.37%	86.66%	80.80%	79.95%
Prefix-Infix(-Suffix) blocking (clean-clean)	96.09%	92.70%	92.70%	97.18%	79.83%	N/A
Prefix-Infix(-Suffix) blocking (dirty)	95.92%	90.33%	88.91%	93.59%	83.50%	N/A
$H3R_2$:						
Token blocking (clean-clean)	96.23%	91.27%	95.62%	88.10%	75.40%	85.64%
Token blocking (dirty)	96.60%	90.16%	94.40%	88.35%	80.67%	N/A ($RR < 0$)
Attribute clustering blocking	97.41%	72.87%	93.38%	80.49%	74.69%	90.66%
Prefix-Infix(-Suffix) blocking (clean-clean)	98.40%	92.11%	89.62%	96.13%	72.40%	N/A
Prefix-Infix(-Suffix) blocking (dirty)	98.33%	89.68%	87.79%	94.73%	77.61%	N/A

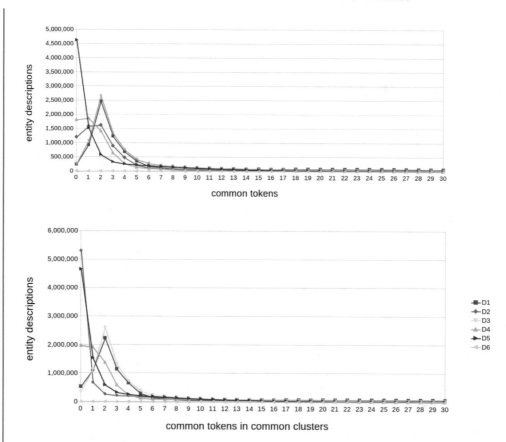

Figure 5.1: Common tokens (top) and common tokens in common clusters (bottom) per entity description distributions for *D1–D6*.

5.3.2   MISSED MATCHES (FNS)

A non-negligible number of matching pairs of descriptions do not share any common tokens at all. Such descriptions, constituting the false negatives of token blocking, should not be assumed faulty, or noisy. We distinguish two different sources of information that can be exploited for successfully placing descriptions of missed matches in common blocks:

i. The matches of their neighbors: Given that an entity description can have, as one of its values, another description, neighborhoods of related descriptions are formed, spinning the Web of data. The knowledge of matches in the neighborhood of a description is valuable for correctly matching this description. For example, if the description *e*10 is related to *e*1, *e*20 is related to $e2$ and we know that $e10$ and $e20$ match, then we can use this knowledge as a hint that $e1$ and $e2$ could possibly match, too.

ii. A third, matching description: In dirty collections (typically peripheral), which are composed of datasets that potentially contain duplicate descriptions, a description $e1$ could have more than one matching description, e.g., both $e2$ and $e3$. Identifying one of these matches, e.g., $(e1, e3)$, knowing that $(e2, e3)$ is a match, leads to also identify the missing match $(e1, e2)$.

Table 5.7 provides details about the number (first row of Table 5.7) and the characteristics of false negative pairs of descriptions, as well as for the set of individual descriptions that constitute these pairs.[11]

We focus first on the neighbors of these descriptions, namely descriptions that appear in their values. We found that almost all the descriptions in the false negatives have at least one neighbor (second row of Table 5.7). Looking more thoroughly, we counted the percentage of descriptions in false negatives that have at least one neighbor belonging to the ground truth (third row of Table 5.7). In all cases, this percentage is more than 10% and goes up to 58% for $D4$. This means that not only do these descriptions have neighbors, but many of these neighbors can be matched to other descriptions in the same collection as well. Then, we counted the percentage of descriptions in false negatives that have neighbors, which have already been matched to another description (fourth row of Table 5.7). This percentage is over 20% in most collections, while it reaches up to 51.84% for $D4$. Finally, we counted the percentage of false negative pairs, whose descriptions have neighbors, which match to each other (fifth row of Table 5.7). This percentage is 0 for $D1$, since matches in this collection are defined as descriptions that have the same subject URI. However, in some peripheral collections ($D2$, $D4$), examining the matches of the neighbors of the descriptions is meaningful. Moreover, in these collections it seems that some specific types of neighbors, e.g., persons and locations, are more likely to be matching than others, e.g., activities. For example, the most frequent predicates that link a match to a pair of matching neighbors in $D2$ are *hasProducer*, *hasDirector*, *starring*, *country* and in $D4$ such predicates are *bandMember*, *associatedBand*, *artist*, *spouseOf*, *siblingOf*.

Another useful piece of information for the missed matches of dirty collections is whether their descriptions have been correctly matched to a third description. The last row of Table 5.7 quantifies this statistic, showing that there are collections, both peripheral ($D2$, $D5$) and central ($D3$), for which this kind of information could, indeed, be useful.

5.3.3   NON-MATCHES (FPS AND TNS)

Next, we examine the ability of blocking methods to identify non-matches, namely their ability to avoid placing non-matching descriptions in the same block. A key statistic for this, regarding the datasets, is the ratio of matches to non-matches, as shown in Table 5.3. The higher the ratio, the easier it is for a blocking method to have better precision, as it statistically has better chances of

---

[11]We have excluded $D6$ from this analysis, as it does not contain any descriptions with neighbors.

Table 5.7: Characteristics of the missed matches (false negatives) of token blocking

	D1	D2	D3	D4	D5
FNs	25,419	3,176	87,672	2,886	303
descriptions in FNs, with neighbor(s)	99.64%	100%	99.99%	100%	100%
descriptions in FNs, with neighbor(s) in ground truth	22.60%	53.94%	36.43%	58.36%	11.57%
descriptions in FNs, with neighbor(s) with an identified match	20.94%	48.54%	34.05%	51.84%	7.59%
FNs with matching neighbors	0%	24.81%	0.38%	37.63%	0%
FNs with common, identified matches	0%	25.35%	10.54%	0.14%	8.58%

suggesting a correct comparison. $D6$ is the collection with the highest such ratio and the highest precision, while $D5$ has the lowest ratio and, in most blocking methods, the lowest precision scores, too. It is clear from Table 5.6 that attribute clustering blocking is the most precise method, since, in almost every case, it results in the fewest wrong suggestions. On the contrary, the worst method in terms of precision is token blocking, in all cases. The differences in precision, in some cases even by an order of magnitude, also determine F-measure, since the differences in recall values are not that big. Note that all the evaluated methods have very low precision, meaning that the vast majority of suggested comparisons correspond to non-matches.

5.4   PERFORMANCE RESULTS

For the benefits of blocking in terms of performance, we examine the number of comparisons that result from the generated blocks and how different this number is to the number of comparisons that we would perform without blocking. Central and peripheral collections do not present different behavior in terms of performance. We consider our $H3R$ and $H3R_2$ measures to be best for an overall evaluation of blocking, since they capture both quality and performance.

Table 5.6 shows that all the evaluated methods manage to greatly reduce the number of comparisons that would be required if blocking was not applied, by one ($D1$–$D4$) or even two ($D5$) orders of magnitude. This is reflected by the high RR scores in all cases. An exception seems to be $D6$, which is much smaller in terms of descriptions and, consequently, comparisons without blocking. Moreover, its descriptions contain many more common tokens than the other collections, leading to more comparisons per entity description. In this case, token blocking does not save many of the comparisons that would be required without blocking and, in the case of seeing $D6$ as a dirty dataset, it even produces twice as many comparisons.

With respect to $H3R$ and $H3R_2$, we notice that, in general, central collections have higher scores, i.e., they present a better balance between recall and reduction ratio. This means that in central collections, the comparisons that are discarded by blocking mostly correspond to non-matches, while many of the comparisons that are discarded by blocking in peripheral collections correspond to matches. Again, $D6$ has a different behavior, since it initially contains a much smaller number of comparisons and a high ratio of matches to non-matches, so the reduction ratio for this collection is limited. Also, these measures are not applicable to token blocking, when applied to $D6$ dirty, since in that case the reduction ratio is negative.

Table 5.8: Recall of token blocking for the collections composed of datasets of Table 5.2 and *BTC12DBpedia*

	airports	airlines	twitter	books	iati		www2012
link	umbel:isLike	umbel:isLike	dct:subject	dct:subject	dct:subject	dct:coverage	foaf:based_near
Recall	97.47%	99.75%	9.52%	63.55%	49.13%	39.46%	62.61%

5.5   DIFFERENT TYPES OF LINKS

In order to evaluate the ability of blocking methods to identify more types of links, semantically close or even not that close to equivalence links, we have run a set of experiments with the peripheral collections consisting of each of the datasets of Table 5.2 and *BTC12DBpedia*. Table 5.8 provides the recall of token blocking, when applied to each of those collections. Similarly to the *owl:sameAs* links, token blocking performs well for links with the semantics of equivalence, as in the *airports* and *airlines* datasets with recall values close to 100%. It also manages to identify many subject associations, as in the cases of *books* and *iati* datasets. It performs poorly in identifying this kind of association, however, in the *twitter* dataset, where its recall values fall to below 10%. This could be justified by the nature of this dataset, which, in most cases, simply states who is the creator of some slides. Regarding spatial associations, token blocking manages to identify a mere 39% of the coverage associations of the *iati* dataset, but it performs much better in identifying the *based_near* associations of *www2012*, with a recall of 63%. The spatial relationships of *coverage* are looser than those of *based_near*, hence the related descriptions are not so strongly related in the former type of links. For example, in *iati*, the description of a project regarding the evaluation of cereal crop residues is linked to the DBpedia resource describing Latin America and the Caribbean, through the *coverage* relation, while, in *www2012*, a Greek professor is linked to the DBpedia resource describing Greece, through the *based_near* relation.

5.6   LESSONS LEARNED

We conclude this chapter with the key points of our evaluation. In general, central entity collections are mostly derived from a common source, Wikipedia, from which they extract information regarding an entity. This way, the descriptions in such collections follow similar naming policies and feature many common tokens (Figure 5.1) in the values of semantically similar, or even equivalent attributes (see the small size of attribute clusters in Table 5.6). Those are exactly the premises on which the evaluated blocking methods are built.

For these reasons, the recall achieved by token blocking in central entity collections is very high (ranging from 94.85% to 99.92%). With the exception of $D6$ (featuring a higher ratio of matching to non-matching descriptions), the precision achieved by token blocking in these collections ranges from $3.64 \cdot 10^{-7}$ to $2.49 \cdot 10^{-6}$. The gains in precision brought by attribute clustering blocking in central entity collections are up to one order of magnitude (for $D3$), with a minor cost on recall (from 0.37% to 3.42%). Prefix-infix(-suffix) blocking can improve both recall and precision of token blocking for central collections, as in $D1$, but it can also deteriorate these values, as in the dirty case of $D3$, which uses random identifiers as URIs, in which recall drops by 7.79% and precision by 26.72%. In a nutshell, many redundant comparisons are suggested by blocking algorithms in all entity collections (see precision and F-measure in Table 5.6), due to the small ratio of matches to non-matches in the collections (Table 5.3). However, as the $H3R$ reveals, the comparisons that are discarded by blocking in central collections mostly correspond to non-matches.

On the other hand, entity descriptions in peripheral datasets are far more thematically diverse, following different naming policies and sharing few common tokens (Figure 5.1), since they stem from various sources. The lack of similar values in those descriptions also leads to a bad clustering of attributes; big clusters, of attributes not similar to each other, are formed (Table 5.6).

For these reasons, the recall of token blocking for peripheral entity collections drops even to 72.13%, while precision ranges from $1.29 \cdot 10^{-9}$ to $1.3 \cdot 10^{-6}$. The gains in precision brought by attribute clustering blocking (up to one order of magnitude) in peripheral collections, come at the cost of a drop in recall up to 24.04% (corresponding to 7,421 more missed matches). Prefix-infix(-suffix) blocking can improve the precision of token blocking in peripheral collections, even by an order of magnitude (for $D4$), or decrease it by an order of magnitude (for $D5$), while it decreases recall from 0.74% to 3.96%, i.e., more matches are missed. In the case of $D4$, in which both datasets use Wikipedia as one of their sources, recall is improved by up to 7.68%. Overall, however, $H3R$ reveals that many of the comparisons that are discarded by blocking in peripheral collections correspond to matches.

Nevertheless, the information that the missed matches provide, e.g., regarding the neighborhoods of their descriptions, as well as other descriptions placed in common blocks with them (Table 5.7), set the ground for a new generation of blocking algorithms, which, unlike the existing ones, will take this information into account and potentially identify those matches, possibly in an iterative fashion.

Finally, in peripheral collections, there are several types of relations, other than equivalence, between the descriptions. Token blocking manages to identify some of these relations, depending on the dataset, the specific type of such links, as well as the immediacy of those relations (Table 5.8). It does not perform well when the data do not contain much information (e.g., see the characteristics of the *twitter* dataset in Table 5.2), or when the relationship of the entities is loose (e.g., see the recall of *iati* for the *coverage* relationship in Table 5.8). Thus, for a quantitative evaluation of blocking methods, ground truth should not be restricted only to *owl:sameAs* links. We could potentially take other relations into account, to identify more such links, or more *owl:sameAs* links, using iterative algorithms.

CHAPTER 6

Conclusions

Although Entity Resolution (ER) has been studied for more than three decades in different computer science communities, it still remains an active area of research. The problem has enjoyed a renaissance during recent years, with the avalanche of data-intensive descriptions of real-world entities provided on the Web by government, scientific, corporate or even user-crafted Knowledge Bases (KBs). To foster an *entity-centric* organization of Web data, it is crucial to reconcile different descriptions, within or across KBs, that refer to the same real-world entity. In this Synthesis lecture, we presented why resolving entities at the Web scale still remains an important and open research problem.

In particular, the *scale*, *diversity*, and *graph structuring* of entity descriptions published according to the Linked Data paradigm challenge the core ER tasks, namely, (i) how descriptions can be effectively compared for similarity and (ii) how resolution algorithms can efficiently filter the candidate description pairs that need to be compared. First, ER in the Web of data involves a large number of KBs (in the order of hundreds) and even a larger number of entity types (in the order of thousands) whose published descriptions could be potentially resolved. In this context, the same entity can be described in a complementary and sometimes conflicting way using highly heterogeneous descriptive schemas. Thus, we need to examine whether two entity descriptions are *somehow* (or near) *similar* without resorting to domain-specific similarity functions and/or mapping rules. This clearly goes beyond deduplication of few collections of descriptions concerning a single entity type and requires leveraging both the content (i.e., literal values of attributes) and the graph structure (i.e., relationships with others) of descriptions. Moreover, in a nutshell, the resolution of some entity descriptions might influence the resolution of others.

Second, since conceiving an ideal similarity function for all kinds of Web data is practically impossible, more pragmatic approaches are needed, that consider different similarity aspects of descriptions at different ER processing steps. We are thus forced to revisit traditional ER workflows consisting of separate indexing (for pruning the number of candidate pairs) and matching (for resolving entity descriptions) phases. As a matter of fact, exact (as blocking) or approximate (as locality-sensitive hashing) indexing techniques, based exclusively on the content of descriptions, fail to achieve a reasonable trade-off between the amount of discarded non-matches and the amount of missed matches. This limitation becomes more severe when resolving entity descriptions published by KBs lying in the center (i.e., with popular entities) and the periphery (i.e., with unpopular, "tail" entities) of the Linked Open Data cloud. Even if it entails an additional processing cost, an iterative entity resolution process seems more promising for minimizing the number of missed matches, while exploring any available intermediate results of blocking and matching in order to discover new candidate description pairs for resolution in an *iterative* manner.

Moreover, the ER problem has been historically framed as a batch, centralized computation aiming to improve data quality in a data warehouse. However, Web applications providing entity-centric search and recommendations strive for new ER execution strategies to resolve, under specific efficiency or effectiveness constraints, very large collections of entity descriptions, eventually arriving in streams. On the one hand, various well-known algorithms have benefited from parallel and distributed implementations aiming to reduce the overall execution time of the entire ER process. On the other hand, the recently proposed progressive ER algorithms interleave the decisions about potential candidate pairs of descriptions for comparisons with partial entity resolution results. Progressive ER aims to identify many similar pairs as early in the detection process as possible and thus have the potential to deliver more complete results in shorter amounts of time compared to a traditional ER execution approach. Still, this new wave of progressive ER is not suited for resolving descriptions that refer to multiple entity types and are graph structured.

Finally, three active research topics that are not covered in this lecture due to space restrictions are probabilistic, privacy-preserving, and crowdsourced-based ER.

- **Probabilistic ER.** The ER approaches presented so far take as input *certain* entity descriptions and produce *certain* matching results. A recent line of research is to consider some confidence scores, i.e., work with *uncertain* data. For example, [Ioannou et al., 2008] assigns a probability to each matching decision, depending on the evidence supporting this decision. Such evidence can be obtained, for instance, by a similarity function. More recently, [Dong et al., 2014a] associates confidence scores with the facts of KBs, representing the belief the fact is correct, based on supervised machine learning methods. Data uncertainty introduces additional challenges to the ER problem, since incommodes the similarities computation and, therefore, the resolution process between entity descriptions; for a review on uncertain entity resolution, readers are referred to [Gal, 2014].

- **Privacy aspects of ER.** The process of ER may raise concerns regarding the privacy protection of individuals, whose descriptions are resolved. Two major issues with respect to privacy, when personal information is matched across organizations, arise: (i) typical ER systems require all data to be available (not only those that are eventually resolved), and (ii) ER results, using entity descriptions from different organizations, can reveal sensitive information that is not available to a single organization [Christen, 2012]. [Whang and Garcia-Molina, 2013] proposes the practice of disinformation, i.e., deliberately injecting false information into the descriptions, in order to protect the privacy of individuals from potential threats posed by ER systems. By adding false information to a description, it becomes less similar to descriptions with which it should match and hence, it is more difficult for an ER system to identify those descriptions as matching. Seen differently, disinformation techniques can be used to evaluate the robustness of an ER system. For a complete survey of privacy-preserving ER, we refer to [Christen, 2012].

- **Crowdsourced-based ER.** In general, crowdsourcing is a costly procedure that can effectively generate a ground truth that will be used either as a training set for a learning-based ER algorithm to identify matches, or to evaluate the results of an ER approach. [Whang et al., 2013a] tries to reduce the cost of crowdsourcing, by minimizing the number of questions that are posed to humans, selecting each time the question with the highest expected accuracy. [Demartini et al., 2013] develops ZenCrowd that uses a semi-automatical ER framework, in which decisions not associated with a high confidence score are propagated to humans. ZenCrowd is also able to identify, and thus ignore, unreliable human decisions. On the opposite side, in [Wang et al., 2012], descriptions are initially resolved by machines and then people only verify the most certain matches, while [Vesdapunt et al., 2014] exploits the transitivity of the equivalence relation to infer as many matches as possible, based on the ER answers that were verified by humans. Existing crowdsourced-based ER approaches are challenged by the size of the datasets in the Web of data [Gokhale et al., 2014].

Bibliography

S. Abiteboul, I. Manolescu, P. Rigaux, M.-C. Rousset, and P. Senellart. *Web Data Management*. Cambridge University Press, New York, NY, USA, 2011. ISBN 1107012430, 9781107012431. DOI: 10.1017/CBO9780511998225. xiv

L. A. Adamic and E. Adar. Friends and neighbors on the web. *Social Networks*, 25(3):211–230, 2003. DOI: 10.1016/S0378-8733(03)00009-1. 30

A. N. Aizawa and K. Oyama. A fast linkage detection scheme for multi-source information integration. In *WIRI*, pages 30–39, 2005. DOI: 10.1109/WIRI.2005.2. 41, 53

Y. Altowim, D. V. Kalashnikov, and S. Mehrotra. Progressive approach to relational entity resolution. *PVLDB*, 7(11):999–1010, 2014. DOI: 10.14778/2732967.2732975. 69, 71

R. Ananthakrishna, S. Chaudhuri, and V. Ganti. Eliminating fuzzy duplicates in data warehouses. In *VLDB*, pages 586–597, 2002. 28, 61

A. Arasu, V. Ganti, and R. Kaushik. Efficient exact set-similarity joins. In *VLDB*, pages 918–929, 2006. 47

S. Araújo, D. Tran, A. DeVries, J. Hidders, and D. Schwabe. SERIMI: class-based disambiguation for effective instance matching over heterogeneous web data. In *WebDB*, pages 25–30, 2012. 24

S. Auer, C. Bizer, G. Kobilarov, J. Lehmann, R. Cyganiak, and Z. G. Ives. Dbpedia: A nucleus for a web of open data. In *ISWC*, pages 722–735, 2007. DOI: 10.1007/978-3-540-76298-0_52. 3

S. Auer, J. Lehmann, and S. Hellmann. Linkedgeodata: Adding a spatial dimension to the web of data. In A. Bernstein, D. Karger, T. Heath, L. Feigenbaum, D. Maynard, E. Motta, and K. Thirunarayan, editors, *ISWC*, volume 5823 of *Lecture Notes in Computer Science*, pages 731–746. Springer Berlin Heidelberg, 2009. 10

N. Augsten and M. H. Böhlen. *Similarity Joins in Relational Database Systems*. Synthesis Lectures on Data Management. Morgan & Claypool Publishers, 2013. ISBN 9781627050289. DOI: 10.2200/S00544ED1V01Y201310DTM038. 26, 47

B. Bahmani, A. Goel, and R. Shinde. Efficient distributed locality sensitive hashing. In *CIKM*, pages 2174–2178, 2012. DOI: 10.1145/2396761.2398596. 38

S. Balakrishnan, A. Y. Halevy, B. Harb, H. Lee, J. Madhavan, A. Rostamizadeh, W. Shen, K. Wilder, F. Wu, and C. Yu. Applying webtables in practice. In *CIDR 2015, Seventh Biennial Conference on Innovative Data Systems Research, Asilomar, CA, USA, January 4-7, 2015, Online Proceedings*, 2015. 7

K. Balog, M. Bron, and M. de Rijke. Category-based query modeling for entity search. In *ECIR*, pages 319–331, 2010a. DOI: 10.1007/978-3-642-12275-0_29. 3, 6, 14

K. Balog, E. Meij, and M. de Rijke. Entity search: Building bridges between two worlds. In *SEMSEARCH*, pages 9:1–9:5, 2010b. DOI: 10.1145/1863879.1863888. 3, 6, 14

R. J. Bayardo, Y. Ma, and R. Srikant. Scaling up all pairs similarity search. In *WWW*, pages 131–140, 2007. DOI: 10.1145/1242572.1242591. 46, 47

Z. Bellahsene, A. Bonifati, and E. Rahm, editors. *Schema Matching and Mapping*. Data-Centric Systems and Applications. Springer, 2011. ISBN 978-3-642-16517-7. URL http://dx.doi.org/10.1007/978-3-642-16518-4. DOI: 10.1007/978-3-642-16518-4. 19

K. Bellare, C. Curino, A. Machanavajihala, P. Mika, M. Rahurkar, and A. Sane. WOO: A scalable and multi-tenant platform for continuous knowledge base synthesis. *PVLDB*, 6(11): 1114–1125, 2013. DOI: 10.14778/2536222.2536236. 68

O. Benjelloun, H. Garcia-Molina, H. Gong, H. Kawai, T. E. Larson, D. Menestrina, and S. Thavisomboon. D-swoosh: A family of algorithms for generic, distributed entity resolution. In *ICDCS*, page 37, 2007. DOI: 10.1109/ICDCS.2007.96. 58

O. Benjelloun, H. Garcia-Molina, D. Menestrina, Q. Su, S. E. Whang, and J. Widom. Swoosh: a generic approach to entity resolution. *VLDB J.*, 18(1):255–276, 2009. DOI: 10.1007/s00778-008-0098-x. 19, 58, 67

L. Bertossi. *Database Repairing and Consistent Query Answering*. Morgan & Claypool Publishers, 2011. ISBN 1608457621, 9781608457625. DOI: 10.2200/S00379ED1V01Y201108DTM020. 5

I. Bhattacharya and L. Getoor. A latent dirichlet model for unsupervised entity resolution. In *SDM*, 2006. 19

I. Bhattacharya and L. Getoor. Collective entity resolution in relational data. *IEEE Trans. Knowl. Data Eng.*, 1(1), 2007. DOI: 10.1145/1217299.1217304. 29, 61

M. Bilenko and R. J. Mooney. Adaptive duplicate detection using learnable string similarity measures. In *Proceedings of the Ninth ACM SIGKDD International Conference on Knowledge Discovery and Data Mining, Washington, DC, USA, August 24 - 27, 2003*, pages 39–48, 2003. DOI: 10.1145/956750.956759. 20

C. Bizer. Search joins with the web. In *ICDT*, page 3, 2014. DOI: 10.5441/002/icdt.2014.04. 6

C. Bizer, J. Lehmann, G. Kobilarov, S. Auer, C. Becker, R. Cyganiak, and S. Hellmann. Dbpedia - A crystallization point for the web of data. *J. Web Sem.*, 7(3):154–165, 2009. DOI: 10.1016/j.websem.2009.07.002. 14

R. Blanco, P. Mika, and S. Vigna. Effective and efficient entity search in RDF data. In *ISWC*, pages 83–97, 2011. DOI: 10.1007/978-3-642-25073-6_6. 3, 6, 14

R. Blanco, B. B. Cambazoglu, P. Mika, and N. Torzec. Entity recommendations in web search. In *ISWC*, pages 33–48, 2013. DOI: 10.1007/978-3-642-41338-4_3. 3, 14

C. Böhm, G. de Melo, F. Naumann, and G. Weikum. LINDA: distributed web-of-data-scale entity matching. In *CIKM*, pages 2104–2108, 2012. DOI: 10.1145/2396761.2398582. 19, 24, 25, 30, 63

K. D. Bollacker, C. Evans, P. Paritosh, T. Sturge, and J. Taylor. Freebase: a collaboratively created graph database for structuring human knowledge. In *SIGMOD*, pages 1247–1250, 2008. DOI: 10.1145/1376616.1376746. 3

A. Bordes and E. Gabrilovich. Constructing and mining web-scale knowledge graphs. In *SIGKDD*, page 1967, 2014. DOI: 10.1145/2623330.2630803. 6

P. Bouquet, H. Stoermer, and D. Giacomuzzi. OKKAM: enabling a web of entities. In *WWW*, 2007. 4, 14

A. Z. Broder. On the resemblance and containment of documents. In *Proceedings of the Compression and Complexity of Sequences*, pages 21–29. IEEE Computer Society, 1997. DOI: 10.1109/SEQUEN.1997.666900. 33

A. Z. Broder. Identifying and filtering near-duplicate documents. In *COM*, pages 1–10, 2000. DOI: 10.1007/3-540-45123-4_1. 5, 18

A. Z. Broder, M. Charikar, A. M. Frieze, and M. Mitzenmacher. Min-wise independent permutations. In *ACM Symposium on the Theory of Computing*, pages 327–336, 1998. DOI: 10.1145/276698.276781. 33

M. J. Cafarella, A. Y. Halevy, D. Z. Wang, E. Wu, and Y. Zhang. Webtables: exploring the power of tables on the web. *PVLDB*, 1(1):538–549, 2008. DOI: 10.14778/1453856.1453916. 6, 28

A. Carlson, J. Betteridge, B. Kisiel, B. Settles, E. R. H. Jr., and T. M. Mitchell. Toward an architecture for never-ending language learning. In *AAAI*, 2010. 4

M. S. Charikar. Similarity estimation techniques from rounding algorithms. In *STOC*, pages 380–388, 2002. DOI: 10.1145/509907.509965. 27

S. Chaudhuri, V. Ganti, and R. Motwani. Robust identification of fuzzy duplicates. In *ICDE*, pages 865–876, 2005. DOI: 10.1109/ICDE.2005.125. 19

S. Chaudhuri, V. Ganti, and R. Kaushik. A primitive operator for similarity joins in data cleaning. In *ICDE*, page 5, 2006. DOI: 10.1109/ICDE.2006.9. 46

S. Chen, B. Ma, and K. Zhang. On the similarity metric and the distance metric. *Theor. Comput. Sci.*, 410(24-25):2365–2376, 2009. DOI: 10.1016/j.tcs.2009.02.023. 23, 24

G. Cheng and Y. Qu. Relatedness between vocabularies on the web of data: A taxonomy and an empirical study. *J. Web Sem.*, 20:1–17, 2013. DOI: 10.1016/j.websem.2013.02.001. 2

P. Christen. Febrl -: an open source data cleaning, deduplication and record linkage system with a graphical user interface. In *SIGKDD*, pages 1065–1068, 2008. DOI: 10.1145/1401890.1402020. 21

P. Christen. *Data Matching - Concepts and Techniques for Record Linkage, Entity Resolution, and Duplicate Detection*. Data-Centric Systems and Applications. Springer, 2012. ISBN 978-3-642-31163-5. DOI: 10.1007/978-3-642-31164-2. xiv, 5, 13, 88

V. Christophides. Resource description framework (rdf) schema (rdfs). In L. LIU and M. ÖZSU, editors, *Encyclopedia of Database Systems*, pages 2425–2428. Springer US, 2009. ISBN 978-0-387-35544-3. 3

K. L. Clarkson. Nearest-neighbor searching and metric space dimensions. *Nearest-neighbor methods for learning and vision: theory and practice*, pages 15–59, 2006. 27

N. Dalvi, R. Kumar, B. Pang, R. Ramakrishnan, A. Tomkins, P. Bohannon, S. Keerthi, and S. Merugu. A web of concepts. In *PODS*, pages 1–12, 2009. DOI: 10.1145/1559795.1559797. 4, 14

N. N. Dalvi, A. Machanavajjhala, and B. Pang. An analysis of structured data on the web. *PVLDB*, 5(7):680–691, 2012. DOI: 10.14778/2180912.2180920. 4, 7, 8, 11, 13, 14

G. de Melo. Not quite the same: Identity constraints for the web of linked data. In *AAAI*, 2013. 20, 21, 24

J. Dean and S. Ghemawat. Mapreduce: simplified data processing on large clusters. *Commun. ACM*, 51(1):107–113, 2008. DOI: 10.1145/1327452.1327492. 41

G. Demartini, D. E. Difallah, and P. Cudré-Mauroux. Large-scale linked data integration using probabilistic reasoning and crowdsourcing. *VLDB J.*, 22(5):665–687, 2013. DOI: 10.1007/s00778-013-0324-z. 89

O. Deshpande, D. S. Lamba, M. Tourn, S. Das, S. Subramaniam, A. Rajaraman, V. Harinarayan, and A. Doan. Building, maintaining, and using knowledge bases: a report from the trenches. In *SIGMOD*, pages 1209–1220, 2013. DOI: 10.1145/2463676.2465297. 4

I. S. Dhillon and D. S. Modha. Concept decompositions for large sparse text data using clustering. *Machine Learning*, 42(1/2):143–175, 2001. DOI: 10.1023/A:1007612920971. 26

A. Doan, J. F. Naughton, R. Ramakrishnan, A. Baid, X. Chai, F. Chen, T. Chen, E. Chu, P. DeRose, B. Gao, C. Gokhale, J. Huang, W. Shen, and B.-Q. Vuong. Information extraction challenges in managing unstructured data. *SIGMOD Rec.*, 37(4):14–20, 2009. DOI: 10.1145/1519103.1519106. 4, 7

A. Doan, A. Y. Halevy, and Z. G. Ives. *Principles of Data Integration*. Morgan Kaufmann, 2012. ISBN 978-0-12-416044-6. xiv, 19

G. Dong. Cross domain similarity mining: research issues and potential applications including supporting research by analogy. *SIGKDD Explorations*, 14(1):43–47, 2012. URL http://do i.acm.org/10.1145/2408736.2408744. DOI: 10.1145/2408736.2408744. 37

X. Dong, A. Y. Halevy, and J. Madhavan. Reference reconciliation in complex information spaces. In *SIGMOD*, pages 85–96, 2005. DOI: 10.1145/1066157.1066168. 19, 65, 69

X. Dong, E. Gabrilovich, G. Heitz, W. Horn, N. Lao, K. Murphy, T. Strohmann, S. Sun, and W. Zhang. Knowledge vault: a web-scale approach to probabilistic knowledge fusion. In *SIGKDD*, pages 601–610, 2014a. DOI: 10.1145/2623330.2623623. 4, 5, 88

X. L. Dong and F. Naumann. Data fusion - resolving data conflicts for integration. *PVLDB*, 2 (2):1654–1655, 2009. DOI: 10.14778/1687553.1687620. 8, 19

X. L. Dong and D. Srivastava. *Big Data Integration*. Synthesis Lectures on Data Management. Morgan & Claypool Publishers, 2015. DOI: 10.2200/S00578ED1V01Y201404DTM040. xiv, 14, 19, 41, 68

X. L. Dong, E. Gabrilovich, G. Heitz, W. Horn, K. Murphy, S. Sun, and W. Zhang. From data fusion to knowledge fusion. *PVLDB*, 7(10):881–892, 2014b. DOI: 10.14778/2732951.2732962. 8, 19

S. Duan, A. Kementsietsidis, K. Srinivas, and O. Udrea. Apples and oranges: a comparison of RDF benchmarks and real RDF datasets. In *SIGMOD*, pages 145–156, 2011. DOI: 10.1145/1989323.1989340. 12, 18

S. Duan, A. Fokoue, O. Hassanzadeh, A. Kementsietsidis, K. Srinivas, and M. J. Ward. Instance-based matching of large ontologies using locality-sensitive hashing. In *ISWC*, pages 49–64, 2012. DOI: 10.1007/978-3-642-35176-1_4. 33

L. Egghe and C. Michel. Construction of weak and strong similarity measures for ordered sets of documents using fuzzy set techniques. *Inf. Process. Manage.*, 39(5):771–807, 2003. DOI: 10.1016/S0306-4573(02)00027-4. 27

O. Etzioni, M. J. Cafarella, D. Downey, A. Popescu, T. Shaked, S. Soderland, D. S. Weld, and A. Yates. Unsupervised named-entity extraction from the web: An experimental study. *Artif. Intell.*, 165(1):91–134, 2005. DOI: 10.1016/j.artint.2005.03.001. 14

J. Euzenat and P. Shvaiko. *Ontology Matching, Second Edition*. Springer, 2013. ISBN 978-3-642-38720-3. DOI: 10.1007/978-3-642-38721-0. 25

A. Fader, S. Soderland, and O. Etzioni. Identifying relations for open information extraction. In *EMNLP*, pages 1535–1545, 2011. 4

L. Fang, A. D. Sarma, C. Yu, and P. Bohannon. REX: explaining relationships between entity pairs. *PVLDB*, 5(3):241–252, 2011. DOI: 10.14778/2078331.2078339. 14

I. P. Fellegi and A. B. Sunter. A theory for record linkage. *Journal of the American Statistical Association*, 64:1183–1210, 1969. DOI: 10.1080/01621459.1969.10501049. 40, 53

A. Ferrara, A. Nikolov, and F. Scharffe. Data linking. *J. Web Sem.*, 23:1, 2013. DOI: 10.4018/978-1-4666-3610-1.ch008. 10

A. Gal. Tutorial: Uncertain entity resolution. *PVLDB*, 7(13):1711–1712, 2014. DOI: 10.14778/2733004.2733068. 88

L. Galarraga, G. Heitz, K. Murphy, and F. M. Suchanek. Canonicalizing open knowledge bases. In *CIKM*, pages 1679–1688, 2014. DOI: 10.1145/2661829.2662073. 60

L. Getoor and A. Machanavajjhala. Entity resolution for big data. In *SIGKDD*, page 1527, 2013. DOI: 10.1145/2487575.2506179. 19

C. Gokhale, S. Das, A. Doan, J. F. Naughton, N. Rampalli, J. W. Shavlik, and X. Zhu. Corleone: hands-off crowdsourcing for entity matching. In *SIGMOD*, pages 601–612, 2014. DOI: 10.1145/2588555.2588576. 89

G. Grahne and J. Zhu. Fast algorithms for frequent itemset mining using fp-trees. *IEEE Trans. Knowl. Data Eng.*, 17(10):1347–1362, 2005. DOI: 10.1109/TKDE.2005.166. 48

L. Gravano, P. G. Ipeirotis, H. V. Jagadish, N. Koudas, S. Muthukrishnan, and D. Srivastava. Approximate string joins in a database (almost) for free. In *VLDB*, pages 491–500, 2001. 40, 53

R. Grishman. Information extraction: Capabilities and challenges. Technical report, NYU CS Dept, 2012. Technical Report. 7

T. Gruber. Collective knowledge systems: Where the social web meets the semantic web. *Web Semantics: Science, Services and Agents on the World Wide Web*, 6(1), 2008. ISSN 1570-8268. DOI: 10.1016/j.websem.2007.11.011. 1

A. Gruenheid, X. L. Dong, and D. Srivastava. Incremental record linkage. *PVLDB*, 7(9):697–708, 2014. DOI: 10.14778/2732939.2732943. 68

C. Guéret, P. Groth, C. Stadler, and J. Lehmann. Linked data quality assessment through network analysis. In *ISWC*, 2011. 10

R. V. Guha. Light at the end of the tunnel - keynote. In *The Semantic Web - ISWC 2013 - 12th International Semantic Web Conference, Sydney, NSW, Australia, October 21-25, 2013, Proceedings, Part I*, 2013. 7

K. Haas, P. Mika, P. Tarjan, and R. Blanco. Enhanced results for web search. In *Proceedings of the 34th International ACM SIGIR Conference on Research and Development in Information Retrieval*, pages 725–734, 2011. DOI: 10.1145/2009916.2010014. 3, 6

O. Hassanzadeh and M. P. Consens. Linked movie data base (triplification challenge report). In *I-SEMANTICS*, pages 194–196, 2008. 10

O. Hassanzadeh, A. Kementsietsidis, L. Lim, R. J. Miller, and M. Wang. A framework for semantic link discovery over relational data. In *CIKM*, pages 1027–1036, 2009. DOI: 10.1145/1645953.1646084. 10

J. He. *Large Scale Nearest Neighbor Search – Theories, Algorithms, and Applications*. PhD thesis, 2014. 5

T. Heath and C. Bizer. *Linked Data: Evolving the Web into a Global Data Space*. Synthesis Lectures on the Semantic Web. Morgan & Claypool Publishers, 2011. DOI: 10.2200/S00334ED1V01Y201102WBE001. 1

M. A. Hernàndez and S. J. Stolfo. The merge/purge problem for large databases. In *SIGMOD*, pages 127–138, 1995. DOI: 10.1145/568271.223807. 41, 53

M. Herschel, F. Naumann, S. Szott, and M. Taubert. Scalable iterative graph duplicate detection. *IEEE Trans. Knowl. Data Eng.*, 24(11):2094–2108, 2012. DOI: 10.1109/TKDE.2011.99. 57

T. Hey, S. Tansley, and K. M. Tolle. Jim gray on escience: a transformed scientific method. In *The Fourth Paradigm: Data-Intensive Scientific Discovery*. 2009. 1

J. Hoffart, F. M. Suchanek, K. Berberich, and G. Weikum. YAGO2: A spatially and temporally enhanced knowledge base from wikipedia. *Artif. Intell.*, 194:28–61, 2013. DOI: 10.1016/j.artint.2012.06.001. 3, 14

A. Hogan, A. Polleres, J. Umbrich, and A. Zimmermann. Some entities are more equal than others: statistical methods to consolidate linked data. In *4th International Workshop on New Forms of Reasoning for the Semantic Web: Scalable and Dynamic (NeFoRS2010)*, 2010. 24

A. Hogan, J. Umbrich, A. Harth, R. Cyganiak, A. Polleres, and S. Decker. An empirical survey of linked data conformance. *Web Semant.*, 14:14–44, 2012. DOI: 10.1016/j.websem.2012.02.001. 12

E. H. Hovy, R. Navigli, and S. P. Ponzetto. Collaboratively built semi-structured content and artificial intelligence: The story so far. *Artif. Intell.*, 194:2–27, 2013. DOI: 10.1016/j.artint.2012.10.002. 4

E. Ioannou, C. Niederée, and W. Nejdl. Probabilistic entity linkage for heterogeneous information spaces. In *CAiSE*, pages 556–570, 2008. DOI: 10.1007/978-3-540-69534-9_41. 88

R. Isele and C. Bizer. Learning expressive linkage rules using genetic programming. *PVLDB*, 5 (11):1638–1649, 2012. DOI: 10.14778/2350229.2350276. 10, 37, 77

R. Isele and C. Bizer. Active learning of expressive linkage rules using genetic programming. *J. Web Sem.*, 23:2–15, 2013. DOI: 10.1016/j.websem.2013.06.001. 10, 77

R. Isele, A. Jentzsch, and C. Bizer. Efficient multidimensional blocking for link discovery without losing recall. In *WebDB*, 2011. 48

D. W. Jacobs, D. Weinshall, and Y. Gdalyahu. Classification with nonmetric distances: Image retrieval and class representation. *IEEE Trans. Pattern Anal. Mach. Intell.*, 22(6):583–600, 2000. DOI: 10.1109/34.862197. 24

E. H. Jacox and H. Samet. Metric space similarity joins. *ACM Trans. Database Syst.*, 33(2), 2008. DOI: 10.1145/1366102.1366104. 27

B. J. Jansen and A. Spink. How are we searching the world wide web? A comparison of nine search engine transaction logs. *Inf. Process. Manage.*, 42(1):248–263, 2006. DOI: 10.1016/j.ipm.2005.06.002. 14

M. A. Jaro. Advances in record-linkage methodology as applied to matching the 1985 census of tampa, florida. *Journal of the American Statistical Association*, 84(406):414–420, 1989. DOI: 10.1080/01621459.1989.10478785. 25

Y. Jiang, G. Li, J. Feng, and W. Li. String similarity joins: An experimental evaluation. *PVLDB*, 7(8):625–636, 2014. DOI: 10.14778/2732296.2732299. 47

Y. Jin, E. Kiciman, K. Wang, and R. Loynd. Entity linking at the tail: sparse signals, unknown entities, and phrase models. In *WSDM*, pages 453–462, 2014. DOI: 10.1145/2556195.2556230. 14

D. V. Kalashnikov and S. Mehrotra. Domain-independent data cleaning via analysis of entity-relationship graph. *ACM Trans. Database Syst.*, 31(2):716–767, 2006. DOI: 10.1145/1138394.1138401. 19

B. Kenig and A. Gal. MFIBlocks: An effective blocking algorithm for entity resolution. *Inf. Syst.*, 38(6):908–926, 2013. DOI: 10.1016/j.is.2012.11.008. 48, 53

H. Kim and D. Lee. HARRA: fast iterative hashed record linkage for large-scale data collections. In *EDBT*, pages 525–536, 2010. DOI: 10.1145/1739041.1739104. 33, 66

I. Kitsos, K. Magoutis, and Y. Tzitzikas. Scalable entity-based summarization of web search results using mapreduce. *Distributed and Parallel Databases*, 32(3):405–446, 2014. DOI: 10.1007/s10619-013-7133-7. 3

L. Kolb, A. Thor, and E. Rahm. Multi-pass sorted neighborhood blocking with mapreduce. *Computer Science - R&D*, 27(1):45–63, 2012a. DOI: 10.1007/s00450-011-0177-x. 41, 53

L. Kolb, A. Thor, and E. Rahm. Dedoop: Efficient deduplication with hadoop. *PVLDB*, 5(12): 1878–1881, 2012b. DOI: 10.14778/2367502.2367527. 41, 53

D. Kontokostas, P. Westphal, S. Auer, S. Hellmann, J. Lehmann, R. Cornelissen, and A. Zaveri. Test-driven evaluation of linked data quality. In *WWW*, pages 747–758, 2014. DOI: 10.1145/2566486.2568002. 4, 8

H. Köpcke, A. Thor, and E. Rahm. Evaluation of entity resolution approaches on real-world match problems. *PVLDB*, 3(1):484–493, 2010. DOI: 10.14778/1920841.1920904. 21

V. I. Levenshtein. Binary codes capable of correcting deletions, insertions and reversals. *Soviet Physics Doklady*, 10:707–710, 1966. 25

T. Lin, P. Pantel, M. Gamon, A. Kannan, and A. Fuxman. Active objects: actions for entity-centric search. In *WWW*, pages 589–598, 2012. DOI: 10.1145/2187836.2187916. 3, 6, 14

G. Malewicz, M. H. Austern, A. J. C. Bik, J. C. Dehnert, I. Horn, N. Leiser, and G. Czajkowski. Pregel: a system for large-scale graph processing. In *SIGMOD*, pages 135–146, 2010. DOI: 10.1145/1807167.1807184. 67

P. Malhotra, P. Agarwal, and G. Shroff. Graph-parallel entity resolution using LSH & IMM. In *EDBT/ICDT Workshops*, pages 41–49, 2014. 67

N. Marie and F. L. Gandon. Survey of linked data based exploration systems. In *ISWC*, 2014. 15

Mausam, M. Schmitz, S. Soderland, R. Bart, and O. Etzioni. Open language learning for information extraction. In *EMNLP-CoNLL*, pages 523–534, 2012. 4

A. McCallum, K. Nigam, and L. H. Ungar. Efficient clustering of high-dimensional data sets with application to reference matching. In *SIGKDD*, pages 169–178, 2000. DOI: 10.1145/347090.347123. 19, 62, 63

R. McCreadie, C. Macdonald, and I. Ounis. Mapreduce indexing strategies: Studying scalability and efficiency. *Inf. Process. Manage.*, 48(5):873–888, 2012. DOI: 10.1016/j.ipm.2010.12.003. 48

A. Metwally and C. Faloutsos. V-smart-join: A scalable mapreduce framework for all-pair similarity joins of multisets and vectors. *PVLDB*, 5(8):704–715, 2012. DOI: 10.14778/2212351.2212353. 48

R. Meusel and H. Paulheim. Heuristics for fixing common errors in deployed schema.org microdata. In *The Semantic Web. Latest Advances and New Domains - 12th European Semantic Web Conference, ESWC 2015, Portoroz, Slovenia, May 31 - June 4, 2015. Proceedings*, pages 152–168, 2015. DOI: 10.1007/978-3-319-18818-8_10. 7

R. Meusel, P. Petrovski, and C. Bizer. The webdatacommons microdata, rdfa and microformat dataset series. In P. Mika, T. Tudorache, A. Bernstein, C. Welty, C. Knoblock, D. Vrandečić, P. Groth, N. Noy, K. Janowicz, and C. Goble, editors, *The Semantic Web – ISWC 2014*, pages 277–292. 2014. ISBN 978-3-319-11963-2. 7, 11

Z. Miklós, N. Bonvin, P. Bouquet, M. Catasta, D. Cordioli, P. Fankhauser, J. Gaugaz, E. Ioannou, H. Koshutanski, and A. Maña. From web data to entities and back. In *CAiSE*, pages 302–316, 2010. DOI: 10.1007/978-3-642-13094-6_25. 4, 14

I. Miliaraki, K. Berberich, R. Gemulla, and S. Zoupanos. Mind the gap: large-scale frequent sequence mining. In *SIGMOD*, pages 797–808, 2013. DOI: 10.1145/2463676.2465285. 48

I. Miliaraki, R. Blanco, and M. Lalmas. From "Selena Gomez" to "Marlon Brando": Understanding explorative entity search. In *WWW*, 2015. DOI: 10.1145/2736277.2741284. 3, 14

Y. Mu and S. Yan. Non-metric locality-sensitive hashing. In *AAAI*, 2010. 24, 38

N. Nakashole, M. Theobald, and G. Weikum. Scalable knowledge harvesting with high precision and high recall. In *WSDM*, pages 227–236, 2011. DOI: 10.1145/1935826.1935869. 4

F. Naumann and M. Herschel. *An Introduction to Duplicate Detection*. Synthesis Lectures on Data Management. Morgan & Claypool Publishers, 2010. DOI: 10.2200/S00262ED1V01Y201003DTM003. xiv, 5, 13, 19

M. Nentwig, M. Hartung, A.-C. N. Ngomo, and E. Rahm. A survey of current link discovery frameworks. *Semantic Web Journal*, 2015. 11

T. Neumann and G. Moerkotte.   Characteristic sets: Accurate cardinality estimation for rdf queries with multiple joins.   In *ICDE*, pages 984–994, 2011. DOI:
10.1109/ICDE.2011.5767868. 18

A.-C. N. Ngomo and S. Auer. Limes: A time-efficient approach for large-scale link discovery
on the web of data. In *IJCAI*, pages 2312–2317, 2011. DOI: 10.5591/978-1-57735-516-
8/IJCAI11-385. 10

F. Niu, C. Zhang, C. Ré, and J. W. Shavlik. Elementary: Large-scale knowledge-base construction via machine learning and statistical inference. *Int. J. Semantic Web Inf. Syst.*, 8(3):42–73,
2012. DOI: 10.4018/jswis.2012070103. 4

L. Otero-Cerdeira, F. J. Rodríguez-Martínez, and A. Gómez-Rodríguez. Ontology matching: A
literature review. *Expert Syst. Appl.*, 42(2):949–971, 2015. DOI: 10.1016/j.eswa.2014.08.032.
13

G. Papadakis, G. Demartini, P. Fankhauser, and P. Kärger.   The missing links: discovering
hidden same-as links among a billion of triples.   In *iiWAS*, pages 453–460, 2010. DOI:
10.1145/1967486.1967557. 45, 49

G. Papadakis, E. Ioannou, C. Niederée, and P. Fankhauser.   Efficient entity resolution
for large heterogeneous information spaces.   In *WSDM*, pages 535–544, 2011a. DOI:
10.1145/1935826.1935903. 42, 50, 53

G. Papadakis, E. Ioannou, C. Niederée, T. Palpanas, and W. Nejdl.   Eliminating the redundancy in blocking-based entity resolution methods. In *JCDL*, pages 85–94, 2011b. DOI:
10.1145/1998076.1998093. 50

G. Papadakis, E. Ioannou, C. Niederée, T. Palpanas, and W. Nejdl. Beyond 100 million entities:
large-scale blocking-based resolution for heterogeneous data. In *WSDM*, pages 53–62, 2012.
DOI: 10.1145/2124295.2124305. 44, 53, 78

G. Papadakis, E. Ioannou, T. Palpanas, C. Niederée, and W. Nejdl. A blocking framework for
entity resolution in highly heterogeneous information spaces. *IEEE Trans. Knowl. Data Eng.*,
25(12):2665–2682, 2013. DOI: 10.1109/TKDE.2012.150. 24, 26, 43, 50, 53, 73, 74, 76

G. Papadakis, G. Koutrika, T. Palpanas, and W. Nejdl.   Meta-blocking: Taking entity res-
olutionto the next level.   *IEEE Trans. Knowl. Data Eng.*, 26(8):1946–1960, 2014a. DOI:
10.1109/TKDE.2013.54. 40, 50, 70

G. Papadakis, G. Papastefanatos, and G. Koutrika. Supervised meta-blocking. *PVLDB*, 7(14):
1929–1940, 2014b. DOI: 10.14778/2733085.2733098. 51

D. Papadias. Nearest neighbor query. In L. LIU and M. ÖZSU, editors, *Encyclopedia of Database
Systems*, pages 1890–1890. Springer US, 2009. ISBN 978-0-387-35544-3. 5, 18

L. Papaleo, N. Pernelle, F. Saïs, and C. Dumont. Logical detection of invalid sameas statements in RDF data. In *Knowledge Engineering and Knowledge Management - 19th International Conference, EKAW 2014, Linköping, Sweden, November 24-28, 2014. Proceedings*, pages 373–384, 2014. DOI: 10.1007/978-3-319-13704-9_29. 10

T. Papenbrock, A. Heise, and F. Naumann. Progressive duplicate detection. *IEEE Trans. Knowl. Data Eng.*, 27(5):1316–1329, 2015. DOI: 10.1109/TKDE.2014.2359666. 70

Y. Raimond, C. Sutton, and M. B. Sandler. Automatic interlinking of music datasets on the semantic web. In *WWW*, 2008. 10

A. Rajaraman and J. D. Ullman. *Mining of Massive Datasets*. Cambridge University Press, New York, NY, USA, 2011. ISBN 1107015359, 9781107015357. DOI: 10.1017/CBO9781139058452. 33, 36, 48

V. Rastogi, N. N. Dalvi, and M. N. Garofalakis. Large-scale collective entity matching. *PVLDB*, 4(4):208–218, 2011. DOI: 10.14778/1938545.1938546. 19, 62

M. Saleem, S. S. Padmanabhuni, A. N. Ngomo, J. S. Almeida, S. Decker, and H. F. Deus. Linked cancer genome atlas database. In *I-SEMANTICS*, pages 129–134, 2013. DOI: 10.1145/2506182.2506200. 8

S. Santini and R. Jain. Similarity measures. *IEEE Trans. Pattern Anal. Mach. Intell.*, 21(9): 871–883, 1999. DOI: 10.1109/34.790428. 24

M. Schmachtenberg, C. Bizer, and H. Paulheim. Adoption of the linked data best practices in different topical domains. In *ISWC*, pages 245–260, 2014. DOI: 10.1007/978-3-319-11964-9_16. 8, 74

A. Shrivastava, T. Malisiewicz, A. Gupta, and A. A. Efros. Data-driven visual similarity for cross-domain image matching. *ACM Trans. Graph.*, 30(6):154, 2011. URL http://doi.acm.org/10.1145/2070781.2024188. DOI: 10.1145/2070781.2024188. 37

P. Shvaiko and J. Euzenat. Ontology matching: State of the art and future challenges. *IEEE Trans. Knowl. Data Eng.*, 25(1):158–176, 2013. DOI: 10.1109/TKDE.2011.253. 13, 18

E. Silva, T. Teixeira, G. Teodoro, and E. Valle. Large-scale distributed locality-sensitive hashing for general metric data. In A. Traina, J. Traina, Caetano, and R. Cordeiro, editors, *Similarity Search and Applications*, volume 8821 of *Lecture Notes in Computer Science*, pages 82–93. Springer International Publishing, 2014. DOI: 10.1007/978-3-319-11988-5. 38

T. Skopal. On fast non-metric similarity search by metric access methods. In *EDBT*, pages 718–736, 2006. DOI: 10.1007/11687238_43. 24

F. M. Suchanek, S. Abiteboul, and P. Senellart. PARIS: probabilistic alignment of relations, instances, and schema. *PVLDB*, 5(3):157–168, 2011. DOI: 10.14778/2078331.2078332. 14

A. Tonon, M. Catasta, G. Demartini, P. Cudré-Mauroux, and K. Aberer. Trank: Ranking entity types using the web of data. In *ISWC*, pages 640–656, 2013. DOI: 10.1007/978-3-642-41335-3_40. 12

A. Tversky. Features of similarity. *Psychological Review*, 84:327–352, 1977. DOI: 10.1037/0033-295X.84.4.327. 24

R. Vernica, M. J. Carey, and C. Li. Efficient parallel set-similarity joins using mapreduce. In *SIGMOD*, pages 495–506, 2010. DOI: 10.1145/1807167.1807222. 49, 53

V. S. Verykios, G. V. Moustakides, and M. G. Elfeky. A bayesian decision model for cost optimal record matching. *VLDB J.*, 12(1):28–40, 2003. DOI: 10.1007/s00778-002-0072-y. 19

N. Vesdapunt, K. Bellare, and N. N. Dalvi. Crowdsourcing algorithms for entity resolution. *PVLDB*, 7(12):1071–1082, 2014. DOI: 10.14778/2732977.2732982. 89

J. Volz, C. Bizer, M. Gaedke, and G. Kobilarov. Discovering and maintaining links on the web of data. In *ISWC*, pages 650–665, 2009. DOI: 10.1007/978-3-642-04930-9_41. 10, 48

J. Wang, G. Li, J. X. Yu, and J. Feng. Entity matching: How similar is similar. *PVLDB*, 4(10): 622–633, 2011. DOI: 10.14778/2021017.2021020. 22

J. Wang, T. Kraska, M. J. Franklin, and J. Feng. Crowder: Crowdsourcing entity resolution. *PVLDB*, 5(11):1483–1494, 2012. DOI: 10.14778/2350229.2350263. 89

M. Weis and F. Naumann. Detecting duplicate objects in XML documents. In *IQIS*, pages 10–19, 2004. DOI: 10.1145/1012453.1012456. 61

M. Weis and F. Naumann. Dogmatix tracks down duplicates in XML. In *SIGMOD*, pages 431–442, 2005. DOI: 10.1145/1066157.1066207. 29

M. Weis and F. Naumann. Detecting duplicates in complex XML data. In *ICDE*, page 109, 2006. DOI: 10.1109/ICDE.2006.49. 61

M. J. Welch, A. Sane, and C. Drome. Fast and accurate incremental entity resolution relative to an entity knowledge base. In *CIKM*, pages 2667–2670, 2012. DOI: 10.1145/2396761.2398719. 68

S. E. Whang and H. Garcia-Molina. Disinformation techniques for entity resolution. In *CIKM*, pages 715–720, 2013. DOI: 10.1145/2505515.2505636. 88

S. E. Whang and H. Garcia-Molina. Incremental entity resolution on rules and data. *VLDB J.*, 23(1):77–102, 2014. DOI: 10.1007/s00778-013-0315-0. 68

S. E. Whang, D. Menestrina, G. Koutrika, M. Theobald, and H. Garcia-Molina. Entity resolution with iterative blocking. In *SIGMOD*, pages 219–232, 2009. DOI: 10.1145/1559845.1559870. 65, 66

S. E. Whang, P. Lofgren, and H. Garcia-Molina. Question selection for crowd entity resolution. *PVLDB*, 6(6):349–360, 2013a. DOI: 10.14778/2536336.2536337. 89

S. E. Whang, D. Marmaros, and H. Garcia-Molina. Pay-as-you-go entity resolution. *IEEE Trans. Knowl. Data Eng.*, 25(5):1111–1124, 2013b. DOI: 10.1109/TKDE.2012.43. 50, 68, 70

W. E. Winkler. The state of record linkage and current research problems. Technical report, Statistical Research Division, U.S. Census Bureau, 1999. 25

C. Xiao, W. Wang, X. Lin, and J. X. Yu. Efficient similarity joins for near duplicate detection. In *WWW*, pages 131–140, 2008. DOI: 10.1145/1367497.1367516. 47, 53

C. Xiao, W. Wang, X. Lin, J. X. Yu, and G. Wang. Efficient similarity joins for near-duplicate detection. *ACM Trans. Database Syst.*, 36(3):15, 2011. DOI: 10.1145/2000824.2000825. 46, 47

S. Yan, D. Lee, M.-Y. Kan, and C. L. Giles. Adaptive sorted neighborhood methods for efficient record linkage. In *JCDL*, pages 185–194, 2007. DOI: 10.1145/1255175.1255213. 41, 53

X. Yu, H. Ma, B. P. Hsu, and J. Han. On building entity recommender systems using user click log and freebase knowledge. In *WSDM*, pages 263–272, 2014. DOI: 10.1145/2556195.2556233. 3, 14

A. Zaveri, A. Maurino, and L.-B. Equille. Web data quality: Current state and new challenges. *Int. J. Semant. Web Inf. Syst.*, 10(2):1–6, 2014. DOI: 10.4018/ijswis.2014040101. 4, 8

C. Zhang, F. Li, and J. Jestes. Efficient parallel knn joins for large data in mapreduce. In *EDBT*, pages 38–49, 2012a. DOI: 10.1145/2247596.2247602. 49

D. Zhang, T. Song, J. He, X. Shi, and Y. Dong. A similarity-oriented RDF graph matching algorithm for ranking linked data. In *CIT*, pages 427–434, 2012b. DOI: 10.1109/CIT.2012.100. 24, 25, 31

Y. Zhen, P. Rai, H. Zha, and L. Carin. Cross-modal similarity learning via pairs, preferences, and active supervision. In *Proceedings of the Twenty-Ninth AAAI Conference on Artificial Intelligence, January 25-30, 2015, Austin, Texas, USA.*, pages 3203–3209, 2015. 37

Authors' Biographies

VASSILIS CHRISTOPHIDES

Vassilis Christophides is a professor of Computer Science at the University of Crete. He has been recently appointed to an advanced research position at INRIA Paris–Rocquencourt. Previously, he worked as a Distinguished Scientist at Technicolor, R&I Center in Paris. He studied Electrical Engineering at the National Technical University of Athens (NTUA), Greece, July 1988, he received his DEA in computer science from the University PARIS VI, June 1992, and his Ph.D. from the Conservatoire National des Arts et Metiers (CNAM) of Paris, October 1996. His main research interests include Databases and Web Information Systems, as well as Big Data Processing and Analysis. He has published over 130 articles in high-quality international conferences, journals, and workshops. He has been scientific coordinator of a number of research projects funded by the European Union, the Greek State, and private foundations on the Semantic Web and Digital Preservation at the Institute of Computer Science of FORTH. He has received the 2004 SIGMOD Test of Time Award and the Best Paper Award at the 2nd and 6th International Semantic Web Conference in 2003 and 2007. He served as General Chair of the joint EDBT/ICDT Conference in 2014 at Athens and as Area Chair for the ICDE "Semi-structured, Web, and Linked Data Management" track in 2016 at Bali, Indonesia.

VASILIS EFTHYMIOU

Vasilis Efthymiou is a Ph.D. candidate at the University of Crete and a member of the Information Systems Laboratory of the Institute of Computer Science at FORTH. The topic of his Ph.D. research is entity resolution in the Web of data. He got his MSc and BSc degrees from the same university in 2012 and 2010, respectively. He has received undergraduate and postgraduate scholarships from FORTH, working in the areas of Semantic Web, non-monotonic reasoning, and Ambient Intelligence.

KOSTAS STEFANIDIS

Kostas Stefanidis is a research scientist at ICS-FORTH, Greece. Previously, he worked as a post-doctoral researcher at the IDI Dept. of NTNU in Norway, with a scholarship funded by the ERCIM Marie Curie Network, and the CSE Dept. of CUHK in Hong Kong. He got his Ph.D. in personalized data management from the University of Ioannina, Greece, in 2009. His research interests lie in the intersection of databases, Web and information retrieval, and include personalized and context-aware data management systems, recommender systems, keyword-based search, and information extraction, resolution, and integration. Kostas has been involved in several international projects and co-authored more than 35 papers in peer-reviewed conferences and journals, including ACM SIGMOD, IEEE ICDE, and ACM TODS. He is the General co-Chair of the Workshop on Exploratory Search in Databases and the Web (ExploreDB), and the Web & Information Chair of SIGMOD/PODS 2016, and the Proceedings Chair of EDBT/ICDT 2016.

ized and context-aware data management systems, recommender systems, keyword-based search, and information extraction, resolution, and integration. Kostas has been involved in several international projects and co-authored more than 35 papers in peer-reviewed conferences and journals, including ACM SIGMOD, IEEE ICDE, and ACM TODS. He is the General co-Chair of the Workshop on Exploratory Search in Databases and the Web (ExploreDB), and the Web & Information Chair of SIGMOD/PODS 2016, and the Proceedings Chair of EDBT/ICDT 2016.

Authors' Biographies

VASSILIS CHRISTOPHIDES

Vassilis Christophides is a professor of Computer Science at the University of Crete. He has been recently appointed to an advanced research position at INRIA Paris–Rocquencourt. Previously, he worked as a Distinguished Scientist at Technicolor, R&I Center in Paris. He studied Electrical Engineering at the National Technical University of Athens (NTUA), Greece, July 1988, he received his DEA in computer science from the University PARIS VI, June 1992, and his Ph.D. from the Conservatoire National des Arts et Metiers (CNAM) of Paris, October 1996. His main research interests include Databases and Web Information Systems, as well as Big Data Processing and Analysis. He has published over 130 articles in high-quality international conferences, journals, and workshops. He has been scientific coordinator of a number of research projects funded by the European Union, the Greek State, and private foundations on the Semantic Web and Digital Preservation at the Institute of Computer Science of FORTH. He has received the 2004 SIGMOD Test of Time Award and the Best Paper Award at the 2nd and 6th International Semantic Web Conference in 2003 and 2007. He served as General Chair of the joint EDBT/ICDT Conference in 2014 at Athens and as Area Chair for the ICDE "Semi-structured, Web, and Linked Data Management" track in 2016 at Bali, Indonesia.

VASILIS EFTHYMIOU

Vasilis Efthymiou is a Ph.D. candidate at the University of Crete and a member of the Information Systems Laboratory of the Institute of Computer Science at FORTH. The topic of his Ph.D. research is entity resolution in the Web of data. He got his MSc and BSc degrees from the same university in 2012 and 2010, respectively. He has received undergraduate and postgraduate scholarships from FORTH, working in the areas of Semantic Web, non-monotonic reasoning, and Ambient Intelligence.

KOSTAS STEFANIDIS

Kostas Stefanidis is a research scientist at ICS-FORTH, Greece. Previously, he worked as a post-doctoral researcher at the IDI Dept. of NTNU in Norway, with a scholarship funded by the ERCIM Marie Curie Network, and the CSE Dept. of CUHK in Hong Kong. He got his Ph.D. in personalized data management from the University of Ioannina, Greece, in 2009. His research interests lie in the intersection of databases, Web and information retrieval, and include personal-

Printed in the United States
by Baker & Taylor Publisher Services